SpringerBriefs in Computer Science

More information about this series at http://www.springer.com/series/10028

Natasha Petrovska • Aleksandar Stevanovic
Borko Furht

Innovative Web
Applications for Analyzing
Traffic Operations

 Springer

Natasha Petrovska
Florida Atlantic University
Boca Raton, FL, USA

Aleksandar Stevanovic
Florida Atlantic University
Boca Raton, FL, USA

Borko Furht
Florida Atlantic University
Boca Raton, FL, USA

ISSN 2191-5768   ISSN 2191-5776   (electronic)
SpringerBriefs in Computer Science
ISBN 978-3-319-33318-2   ISBN 978-3-319-33319-9   (eBook)
DOI 10.1007/978-3-319-33319-9

Library of Congress Control Number: 2016942565

The use of "Google Maps™" refers to the Google Maps™ mapping service, a registered trademark of
Google Inc.

Printed on acid-free paper

This Springer imprint is published by Springer Nature
The registered company is Springer International Publishing AG Switzerland

Preface

The road traffic along with other key infrastructure sectors such as telecommunication, power, and others has an important role in the economic and technological growth of one country. Traffic engineers and analysts are responsible for solving a diversity of traffic problems, such as traffic data acquisition and evaluation. In response to the need to improve traffic operation, researchers implement advanced technologies and integration of systems and data and develop state-of-the-art applications. This book introduces three novel web applications with an aim to offer traffic operators, managers, and analysts the possibility to monitor the congestion and analyze incidents and signal performance measures. They offer more detailed analysis providing users with insights from different levels and perspectives. The benefit of providing these visualization tools is more efficient estimation of the performance of local transport networks, thus facilitating the decision-making process in case of emergency events.

Boca Raton, FL

Natasha Petrovska
Aleksandar Stevanovic
Borko Furht

Contents

1   Introduction 1

2   Background and Related Work 3
    2.1   Traffic Congestion 3
    2.2   Methodologies for Estimating Congestion Level 4
    2.3   Intelligent Transportation Systems 5
    2.4   Traffic Data Visualization and Visualization Tools 8
    2.5   Maps as Traffic Visualization Tools 12
    2.6   Digital Map Processing 20

3   Visualization Tools for Traffic Congestion Estimation 23
    3.1   Introduction 23
    3.2   Visual Analytics 24
    3.3   Architecture 24
    3.4   Visualization 26
    3.5   Methods to Quantify Road Traffic Congestion 27
    3.6   Demonstration of the Application 28

4   Web Application for Incident Analysis 33
    4.1   Introduction 33
    4.2   Nature of Incident Data 34
    4.3   Concept of the Application 34
    4.4   User Interface 35
    4.5   Used Technology 38

5   Web Application for Measuring Travel-Based
    Time Signal Performance 39
    5.1   Introduction 39
    5.2   System Components 40
    5.3   Data Collection 40

5.4   Estimating Signal Performance...   41
      5.4.1   Volume-Delay Functions...   41
      5.4.2   Estimating Volume/Capacity Ratio.......................   41
      5.4.3   Estimating Level of Service..................................   43
      5.4.4   Estimation of Number of Cycles to Pass
              Through the Signal..   43
5.5   Visualization of VDF Measures...   43
5.6   Implemented Technology..   46

6   Conclusion   47

References   49

Chapter 1
Introduction

Road transportation has big influence on economy, social life and environment of every country. Growing complexity of urban transport networks creates a challenge in managing transport operation in the cities. Problems with traffic monitoring, estimation and control are often solved with the help of Intelligent Transportation Systems (ITSs) and Information and Communication Technologies (ICTs). ITSs combine advanced technologies using real-time information regarding roadway conditions in order to apply appropriate strategies for control.

Transportation systems are being observed at a remarkable extent which results in huge traffic and incident databases. The issue of road traffic congestion is important and ubiquitous and the integration of different technologies and systems helps enormously in solving this issue. From economic perspective the importance can be noticed from the following example, 1 % of the European Union's GDP in 2010 (2 % in 2009) was dedicated to reducing the congestion. Also, 1.3 million accidents per year are the reason of more than 40,000 deaths and 1.7 million of injuries on the roads. Nearly 24 % of the driving time is spent in traffic jams or waiting in stop lights.

Improperly operated traffic lights produce delays that decrease productivity, waste fuel, pollute the air and increase the level of stress. Therefore, conventional fixed-phase signal control is imperfect. Furthermore, the positioning and maintenance of road traffic signs, particularly those which use text to express meaning, has very high costs. For example, in the United States, each road sign costs approximately between $100 and $500 in order to be replaced.

The traffic control through an adequate location of devices can: reduce the *delay* for the traffic, decrease the occurrence of *collision*, optimize *management* of individual intersections, improve *traffic flow*, mitigate the *traffic congestion* in urban areas by reducing the *total number of vehicles* on the road, increasing the *existing road infrastructure*, and increase the *throughput* of already existing infrastructure.

© The Author(s) 2016
N. Petrovska et al., *Innovative Web Applications for Analyzing Traffic Operations*, SpringerBriefs in Computer Science,
DOI 10.1007/978-3-319-33319-9_1

As stated by World Bank (2001) the pace of annual growth of people possessing vehicles is between 15 and 20 % in the developing world. This growth in traffic implies rise in the amount of traffic sent via ITSs and continually impedes the increase of road space in developing countries which consequentially obstructs the efficient functioning of the urban economics in these countries.

As the traffic information sent via ITSs increases rapidly, and because of the need enormous traffic databases to be efficiently used by the traffic operators and managers, developing innovative applications and state-of-the-art visualization tools is indispensable. Visualization provides extraction of data of interest by presenting them in various visual forms using various interactions.

In this book, we introduce three novel Web applications which are good visualization tools that could offer a number of benefits to the public and authorities in decision making processes.

The first application [69] is aimed to help traffic operators in traffic congestion estimation. It visualizes different levels of congestion using the Google Maps™ traffic layer. Different levels of congestion are calculated, designated with green, yellow, red, and gray, correspondingly to normal, medium, high and severe congestion, expressed in %, and updated automatically with the time frequency chosen by the user. The results of the estimated congestion levels are saved in output files in text formats.

The second application is intended to make the incidents databases more comprehensible for the traffic operators and analysts by offering tabular and graphical visualization on Google Maps™ as well as different filtering possibilities.

The third application is used to express signal performance measures such as V/C, Level of Service, and number of cycles waited on intersection by using the vehicles travel time. It offers two formulas for calculating these measures and proposes manual or automatic visualization in table or graphically on Google Maps™. When graphical representation on Google Maps™ is chosen, the calculated measures are presented with Info windows on the map and updated periodically within a time period designated by the user. It uses input files in text format with travel time data.

The book is organized as follows: the Chap. 2 presents research of what has been done in the related fields and technologies such as Intelligent Transportation Systems, visualization tools implemented in transportation systems monitoring, maps like visualization tools, and variety of applications developed for enabling traffic data analysis and visualization. The Chap. 3 describes the components, interface and operation of the application for congestion level estimation, the Chap. 4 demonstrates how the application for incidents analysis works, and the Chap. 5 explains the technologies, interface and how the application for travel time based signal performance measures functions. The Chap. 6 gives the summary of the presented material, contributions and a proposal for potential future improvements.

Chapter 2
Background and Related Work

Vehicle traffic has been an important research issue. Numerous studies have been conducted providing insights from various levels and perspectives. Researchers analyze traffic in terms of speed, flow rate, density, volume, occupancy, congestion, etc.

Traffic *density* is the number of vehicles within unit length of a roadway. This parameter is a measure of effectiveness of many traffic systems. *Occupancy* is a variable which is an indicator of traffic density. It is the fraction of time, out of the total observation period, within which the location where measurement was performed was occupied by vehicles. *Level of Service* is a qualitative measure that expresses traffic situations in terms of travel time, speed, safety, freedom to maneuver, convenience, traffic interruptions etc. There are six categories of LOS labelled by letters A (the best condition) to F (heavily congested flow with traffic demand exceeding the road capacity) [1]. *Capacity* is the maximum theoretical traffic flow rate that a road section is able to accomplish under a certain set of environment, road and traffic conditions. A *number of cycles* waited to pass through a signal is a good indicator of congestion on a signalized approach. Sometimes this measure offers improved sense for congestion ahead than the measured travel time for both, drivers and traffic operators.

The work related to ours covers several areas of research: *traffic congestion and congestion level estimation, Intelligent Transport Systems, traffic data visualization, maps as visualization tools*, and *digital map processing*.

2.1 Traffic Congestion

Generally, traffic can be "free" or "congested". The issue of road traffic congestion is important and pervasive. As a phenomenon, congestion is defined in many different ways. Morris J. Rothenberg defines urban highway congestion as "a condition

© The Author(s) 2016
N. Petrovska et al., *Innovative Web Applications for Analyzing Traffic Operations*, SpringerBriefs in Computer Science,
DOI 10.1007/978-3-319-33319-9_2

in which the number of vehicles attempting to use a roadway at any given time exceeds the ability of the roadway to carry the load at generally acceptable service levels" [2]. Another definition is given by Zhicai et al. [3]. According to them, congestion is a condition which emerges when there is a great reduction in the average velocity (with respect to the free flow), rise in travel time, stop-and-go of vehicle (in addition to increased and frequent velocity variety) as well as when increase in density appear. They mentioned two types of congestion, driven by demand and driven by supply, and studied the socio-economic impact.

2.2   Methodologies for Estimating Congestion Level

Some of the factors that cause traffic congestion are accident, construction work, bad weather and poor traffic signal timing. There are various methodologies for estimating congestion level. Thianniwet et al. [4] proposed a method with 91.29 % accuracy for congestion level measurement using *vehicle movement patterns*. The vehicles' velocities were collected from participants' GPS devices through mobile data networks. The data were sent to the learning model, processed and represented using *sliding windows technique* and *decision tree*. The first technique is used to capture the consecutive moving average velocities, called moving patterns. Human judgments were used to categorize the traffic congestion levels into: light, heavy, and were highly consistent with the results gained by the authors.

Pattara-Atikom, Pongpaibool and Thajchayapong [5] categorized three levels of congestion on main roads in Bangkok utilizing the *weighted exponential moving averages of vehicle speed* gained from GPS data and implementing *threshold technique*. Data were gained from GPS device in a vehicle, and image processing video camera mounted on a vehicle's dashboard. The measured average velocities were grouped into three congestion levels, i.e., red, yellow, and green. The algorithm has error of around one third of a congestion score with 35.29 % over-estimation, 17.65 % underestimation, and 47.066 % correct estimation. There are several benefit of this technique for congestion level classification. Comparing to existing techniques it is comprehensible, straightforward, with minimum requirement of input data (merely vehicle velocity), being measurable and appropriate to automated report, and being compatible with existing traffic report systems in Bangkok.

Traffic congestion level can also be determined using different parameters like *travel time, travel time index, travel delay, travel ratio index, speed reduction index,* etc. In concept of Level of Service, it is considered that congestion occurs when V/C (volume over capacity ratio) exceeds a certain threshold. B. S. Kemer categorized traffic pattern into *free* and *synchronized flow*, and *wide moving jam* [6]. Travel Time Index is expressed as the ratio of real travel time to free flow travel time [7].

J. Lu and L. Cao [8], and Pongpaibool et al. [9] used *fuzzy logic* as a tool for congestion level estimation. J. Lu and L. Cao evaluated congestion from *traffic flow* information based on *adaptive neuro-fuzzy inference system* and series of fuzzy logic rules. They consider level of congestion as a continuous variable changing from free flow to traffic bottleneck. Pongpaibool et al. designed a system which identifies three levels of traffic congestion in Bangkok from data gained via *image processing* using manually set *fuzzy logic* and adaptive *neuro-fuzzy techniques*. As inputs for the fuzzy logic technique they used traffic *volume* and *velocity*, as well as human estimation of the congestion level. The traffic data acquisition was performed utilizing vehicle detection and tracking software, which took a signal from the road-traffic cameras. The results they achieved had 88.79 % accuracy or manually tuned fuzzy logic, and 75.43 % accuracy when adaptive neuro-fuzzy technique was applied.

Combining various factors such as speed, volume, density, travel time, etc. can be another way for quantifying congestion. One of the traffic parameters which can be directly implemented for determining the level of congestion is *traffic density*. There are various historic and real time methods being used by researchers for short term prediction of traffic density including, statistical methods, machine learning techniques and model based techniques.

One way to categorize the approaches for congestion level estimation is by: using **historical** traffic data or using **real-time** data. ExTrac system proposed by Damaiyanti et al. [10] extracts traffic congestion data from historical traffic data and answers the queries about them. The problem with enormous amount of collected historical traffic data (which usually occurs when the patterns are obtained from great quantity of historical traffic data with different features) they solved by creating an efficient data structure model of traffic congestion patterns in spatio-temporal heatmap[1]. They converted collected speed data into congestion degree values, created matrices and summarized them by applying MapReduce framework. The authors used NoSQL database for storing the traffic patterns.

Padiath et al. [11] used *historic and Artificial Neural Network technique* for predicting traffic density on roads in Chennai. Data for this study were gathered using videographic technique from a 1 km three lane road in the Rajiv Gandhi, Chennai, India. Pongnumkul et al. [12] applied historical traffic data patterns and rendered them spatially in a *grid*. Diker and Nasibov used *historical traffic patterns* of the road segments and clustered them according to road traffic congestion [13]. The technique they used was *Fuzzy Neighborhood Density-Based Spatial Clustering of Applications with Noise (FN-DBSCAN)*.

2.3   Intelligent Transportation Systems

Congestion can be referred as traffic demand exceeding the road infrastructure capacity. Since adding supplementary roadway infrastructure to meet the rising population possessing vehicles is limited, and increasing the infrastructure will not correspond to the increase in traffic demand, congestion issue is progressively

---

[1] Heatmaps are a method for representing spatial data that identifies the high-occurrence regions without complicating the overall view.

solved by implementing Intelligent Transportation System (ITS) [11]. ITS control strategies include estimating flow on roadways, dynamically retiming traffic signals, handling traffic incidents, and conveying information to travelers about travel conditions, alternative routes, and alternate modes [70, 71]. Its aim is to diminish traffic load, improve safety, economic vitality, transportation efficiency, and environmental pollution. The system comprises advanced data transfer and communication technology, wide diversity of static and dynamic sensors, and computer control technology. The core of ITS is the acquisition of traffic related data in order to estimate the traffic. Besides traffic management, it also offers traffic reporting to advice travelers. System implements vehicle-to-vehicle (V2V) and vehicle-to-infrastructure (V2I) (vehicle-to-road V2R) communication. Examples of major applications are TrafficCast using BlueTOADs, Palm Beach County, Florida [14]; RITIS [15] using cameras, in USA; Signal Performance Metrics (UDOT) [16]; TRANSGUID in San Antonio, United States; a road traffic information communication system (VIC) in Tokyo, Japan; Electric Road Pricing (ERP) in Singapore; ROMANSE system in Southampton United Kingdom and Germany with its FCD system.

Traffic monitoring and control problems are extensively solved with Intelligent Transportation Systems (ITSs), as well as Information and Communication Technologies (ICTs). With an intention to facilitate assessment of signal operations quality, researchers have developed various data performance measures.

Balke et al. [17] developed a system, called the Traffic Signal Performance Monitoring System (or TSPMS for short) in order to obtain data from both, the traffic signal system, and the detection system to generate traffic performance measures in real time. The performance measures they generated were: cycle time, time to service, queue service time, interval duration, number of vehicles entering per interval, yellow and all-red violation rates, and phase failure rate.

Highway Capacity Manual 2010 [18] supplies the methods for estimating the performance of a signalized intersection which include computing V/C, saturation flow rate, delay, and three determining LOS. Liu et al. [19] introduced a system, named SMART-SIGNAL (Systematic Monitoring of Arterial Road Traffic Signals) which collects and archive event-based traffic signal data on 11-intersection corridor along France Avenue in Minneapolis, MN. Utilizing these data,. SMART-SIGNAL can generate time-dependent performance measures such as intersection queue length and arterial travel time for intersections as well as arterials. The empirical results which authors gained for these measures are greatly consistent with the observed data.

Bezuidenhout et al. [20] offered a new model for queue prediction which relied on the data obtained from a loop detector set at the stop line of signalized intersection. Zaiat et al. [21] introduced intelligent transport system which is potentially very beneficial tool and can help authorities and other decision makers to facilitate transport monitoring and management. It can supply the required information for strategic planning, particularly in case of emergency events. The authors' approach allows Level of Service (LOS) analysis for different transportation modes at geographic abstraction of three levels, from entire monitored region to its sub-areas.

They performed test using input data from road sensors encompassing traffic volume, average speed and lanes occupancy rate. Then using the road LOS calculation methods, they selected one of these values and attributed to a letter from "A" to "F", designating road LOS.

Sen et al. [22] envisioned a system where the used technique includes a pair of road side acoustic sensors. They used Doppler shift of the honk frequency to estimate the vehicle's speed, and thus to calculate road congestion. Using more than 18 h of road-side recordings, they demonstrated that the technique is effective in real conditions. By implementing the vehicle speed distribution, and the number and duration of honks, authors also classified traffic state as either free-flowing or congested with a threshold-based categorization correctness of 70–100 %.

Most travel information systems work by acquiring data from a variety of sensors such as vehicle tracking units, loop detectors, camera, infrared detectors, RFID tags, laser scanners, ultrasonic, radar, GPS tracking units and cellphones. These data are processed centrally and updates are transferred to travelers using a variety of channels. Hence, the main traffic information comes from various transportation management departments and part of the transportation business and is basically announced to meet their own need. In contrast, the dynamic transportation information aimed for the general public seems to be less. Many drivers disregard these data, considering them obsolete or not applicable. Instead of using these information, researchers started to gain travel information automatically from current road users from the mobile network or users' mobile devices. They found that these data have advantages over data collated via sensors.

Vehicular ad-hoc networks (VANETs) is very important segment of ITS. Vehicular ad hoc network (VANET), is the state of the art technology that is part of wireless ad hoc networks which may use vehicles as mobile nodes. The advantage of VANET is rapidly changeable topology which allows dynamic route planning. This can prevent clogging of vehicles in bad circumstances like accidents, fog, etc. by providing alert messages and warnings. As a component of Intelligent Transport Systems (ITSs) its main objective is to provide safety, comfort in traffic, improve decision making processes and introduce innovative effective real-time information for the drivers.

There are mainly two types of communication in VANET: Inter-Vehicular (V2V) and vehicle-to-infrastructure (V2I) or vehicle-to-Roadside Unit (V2R) wireless communication.

A promising new approach for diminishing congestion, is *Virtual Traffic Lights* (VTL). It is introduced with an intention to replace the physical traffic signals and convey virtual traffic information to the driver inside the car in a manner which won't compromise driving security [23, 24]. This is achieved by using technology mounted inside the cars which enable drivers to self-organize on the intersections. Such organization also implies decrease of the urban traffic congestion issue. The technology implemented is V2V communications and DSRC.

Nakamurakare et al. [25] implemented VTL system on Android based mobile phones. Without any help of the intersection, merely by using V2V communication, conflicts on intersections are detected and resolved in ad hoc manner. It makes use of and operates based on the input provided by the built-in modules such as GPS, Map database, and NTP server (for time synchronization). Two different map formats are used in the implementation: XML format used in VTL module and Google Map™ format used for display purpose. The simulation results they got have proved that the VTL can improve the traffic flow up to 60 %.

2.4 Traffic Data Visualization and Visualization Tools

The big amount of traffic data stored in transportation databases impedes human understanding and extracting traffic patterns from these data directly. Taking into consideration the heterogeneity of such complex and large data such as transportation data, data visualization is substantial and inevitable for their analysis. It alleviates the process of detecting the structure, features, anomalies, patterns, and inter-connections in complex data, which is generally complicated. Visualization stipulates various visual forms and different interactions for data representation. It not only offers a qualitative overview of complex data sets, but it can also facilitate the identification of areas of interest and parameters for more specific quantitative analysis. This motivated some researchers to focus on creating visualization tools to assist human in understanding traffic patterns.

Shekhar et al. [26] introduced a web-based visualization package (CubeView) for monitoring sensor network measurements collected from the freeway system in Minneapolis—St. Paul (Twin-Cities) metropolitan area. The system is developed for the Minnesota Department of transportation (MNDOT). It offers extraction of patterns and rules from the historical data to support decision making. The sensor network comprises approximately 900 stations. Depending on the number of lanes, sensors incorporate one to four loop detectors. Sensors observe the *occupancy* and *volume* of traffic on the road and these data are sent to the Traffic Management Center. In CubeView, raw data obtained by loop detectors are saved in binary format, converted into text data and later stored into database servers. After querying the database, data are represented using *clusters*, *classes* and *cube view*. Patterns are rendered using *graph*, *charts*, and *detector maps*. Congestion can also be visualized. The transportation visualization tools can be used by traffic managers, traffic engineers, travelers and commuters, as well as researchers and planners.

Piringer et al. [27] studied the surveillance videos in a tunnel. They automatically detected and prioritized different types of events and marked them in space and time. For each event, users could check the original videos. Zaiat's et al. [21] approach includes several visualization tools: a *map-based view* on the local transport systems performance state; filtering the information on the dashboard by transport domains, modes, components; aggregations for any LOS and geographic abstraction; and *charts view* of the transport system behavior over time.

Lu et al. [28] proposed the Advanced Interactive Traffic Visualization System (AITVS) which presents data cube visualization tools for real-time and historical pattern analysis. It is a web-based visualization system introducing innovative visualization components like spatial and temporal *plots* and *data cube* to analyze and monitor traffic conditions, volume, speed, and occupancy, and thus to overcome the disadvantages of other existing systems. AITVS is more oriented to traffic analysts and managers then to travelers, similar to our proposed web applications.

Pack [29] proposed a web-based visual analytics system for observing and identifying main bottlenecks. It implements variety of visualization tools: a dashboard with map and a popup window for displaying the travel time index, different contour plots, interactive animated map for showing average speeds, travel times, reliability etc., interactive charts and graphs, and performance summary table. As part of the bottleneck ranking very interesting visualization tool is the time spiral graphic which shows the time of occurrence and how long the bottleneck became until it ended (Fig. 2.1 adopted from [29]).

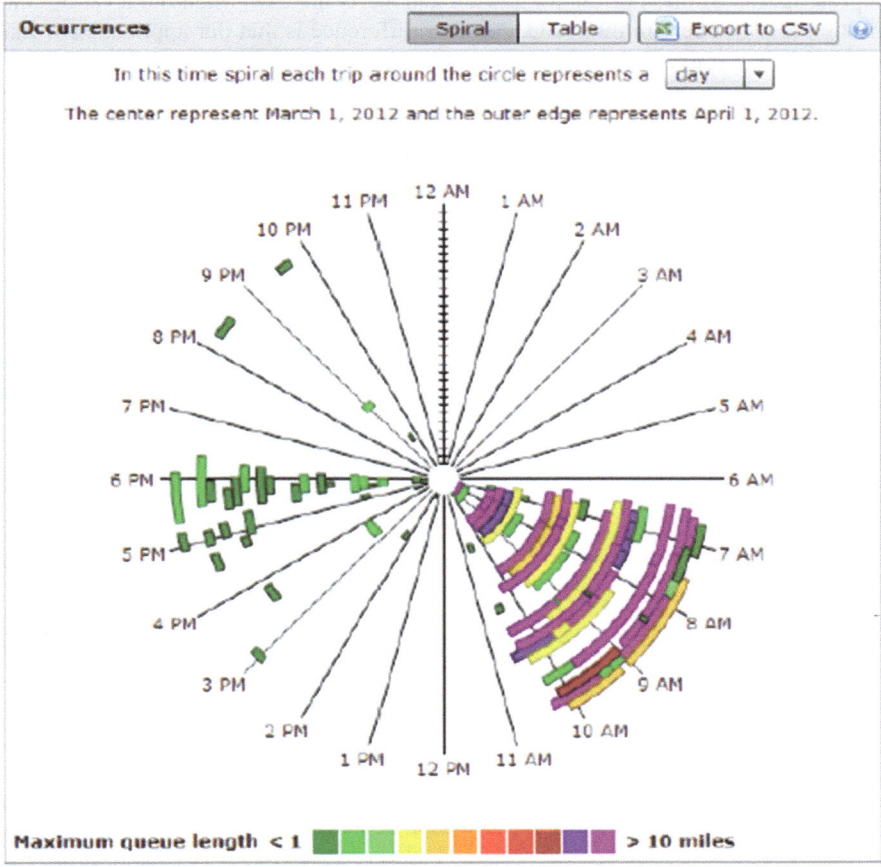

Fig. 2.1 Time spiral graphic showing when a bottleneck occurred and how long it became during that time (Adopted from [29])

Generally, Web-based traffic visualization tools are easily accessible and can be used for simplifying complex and monotonous statistical data providing beneficial information to traffic professionals as well as travelers.

Numerous visualization tools have been implemented specifically for *congestion level analysis and interpretation*. An example for *congestion level estimation and representation* is CongestionGrid [12]. It is a platform which automatically acquires current congestion data from a traffic data provider and enables users to visualize the historical patterns by displaying them in a *grid*. CongestionGrid allows users to explore the temporal traffic patterns by viewing the congestion data of a certain week or an aggregation of data over a period of time. This approach for visual representation of the traffic states is very similar to visual representation used in our web application for congestion estimation, implementing red, yellow, and green colored cells which correspond to high, normal and low traffic, respectively (Fig. 2.2 adopted from [12]). The advantage of our application is that it offers more precise representation of the level of congestion and alerts the user when the level of congestion is above certain threshold. Another difference is that our application works with real-time traffic data.

Diker and Nasibov [13] and Yoon et al. [30] used *maps*, *graphs* and *clusters* as visualization tools. Diker and Nasibov arranged road segments according to certain traffic congestion level into clusters, and Yoon, besides threshold-based quadrant clustering, used *spatio-temporal traffic status plots* of trace data.

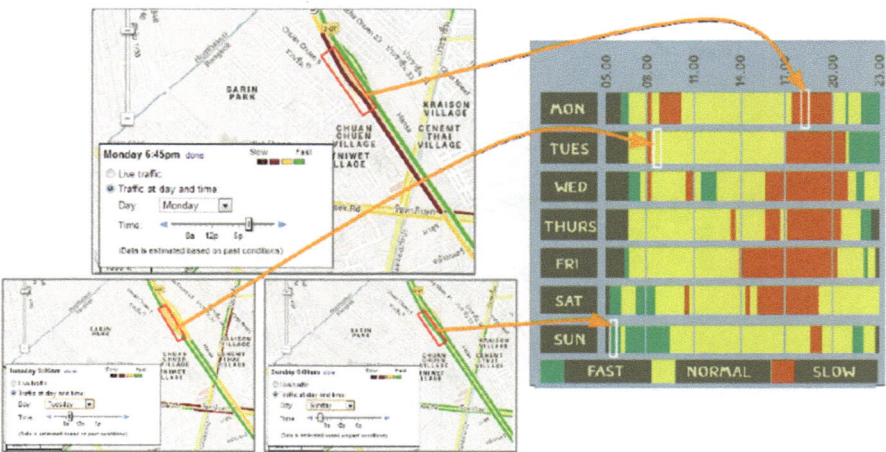

Fig. 2.2 Manual creation of CongestionGrid from Google Maps™'s historical traffic data. CongestionGrid automatically collects current congestion data from a traffic information provider (Google Maps™ Traffic Layer) and allows users to explore and visualize the historical weekly patterns of traffic data, or an aggregation of data over a period of time, by spatially arranging them in a grid. The traffic states are denoted by *red, yellow,* and *green* colors, representing high, normal and low traffic, respectively. (Adopted from [12]). *Map © 2015 Google Inc, used with permission. Google and the Google logo are registered trademarks of Google Inc.*

Wang et al. [31] utilized traffic trajectories (as major type of traffic data received from road sensors) as well as incidents, road speed, and traffic jams as a source for visualization. This model implemented in Beijing also structures the relationships between traffic jams. Traffic trajectories visual analysis most often requires *aggregation* such as *density map* [32]. Density map plots the trajectory density and enables "hot" spots detection. Some of the techniques that authors used are: *propagation graph* level estimation, traffic jam density *visualization on a map* (OpenStreetMap), *topological filters, temporal and size filters, map matching* etc. In order to visualize propagation graph, authors applied *animation, flow map* and *graph layout techniques*. The system introduces five views for visualization: (1) pixel-based road velocity view, to show the speeds and events; (2) graph list view for displaying propagation graphs; and (3) the graph projection view to render propagation graphs' topological relationships; (4) spatial view to represent the traffic congestion density, as well as the propagation path of one designated graph; and (5) multi-faceted filter view, which offers filtering by time, space, size and topology, and through sorting by size and similarity presents structured representation of the propagation graphs. First they estimated free flow speed for every road segment, and by considering relatively low road-velocity they automatically detected traffic congestion events at roads. The interaction of these events in propagation graphs depicted the spatial and temporal propagation of traffic jam.

Unlike this approach, Pack et al. [33] and Khotanzad [59] instead of traffic trajectories, analyzed transportation incident datasets. They designed web-based, visual analytics tool called *Incident Cluster Explorer*. It is an application that provides an integrated view interface to illustrate the spatial, temporal and multidimensional features of the incidents. Users' interaction is supported by including options for selection, filtering and incidents clustering, and accordingly focus on a smaller dataset. Simultaneous interactions between multiple visualization tools are possible, such as *histograms, interactive maps, two-dimensional and parallel coordinates plots*. Figure 2.3 (adopted from [33]) displays an interactive histogram for chosen variable "incident type" in the data set along with an icon mode map representation.

To visualize relationships between a selected pair of variables authors used either *scatter plot mode* or *grid mode*. Also, two mapping modes are allowed: *icon mode* and *heat mode*. In contrast to some websites (FARS) which presents a substantial row data and do not propose any visualization possibility, leaving this complicated task to the user who can only download them, this application is much more comprehensive, sophisticated and user-friendly. Another advantage of this tool over other commercial data visualization products such as Spotfire [34, 62] and Tableau [35, 63] is using the *heat maps* in order to render data on a map to avoid occlusion and overcrowding when it comes to large datasets such as transportation incident data.

Anwar et al. [36, 68] introduced a straightforward method called *Traffic Origins*, to visualize the impact road incidents have on congestion and on vehicle flow in their vicinity, as well as the cascading influence multiple incidents may have on a road network. The incident location is designated with an expanding circle

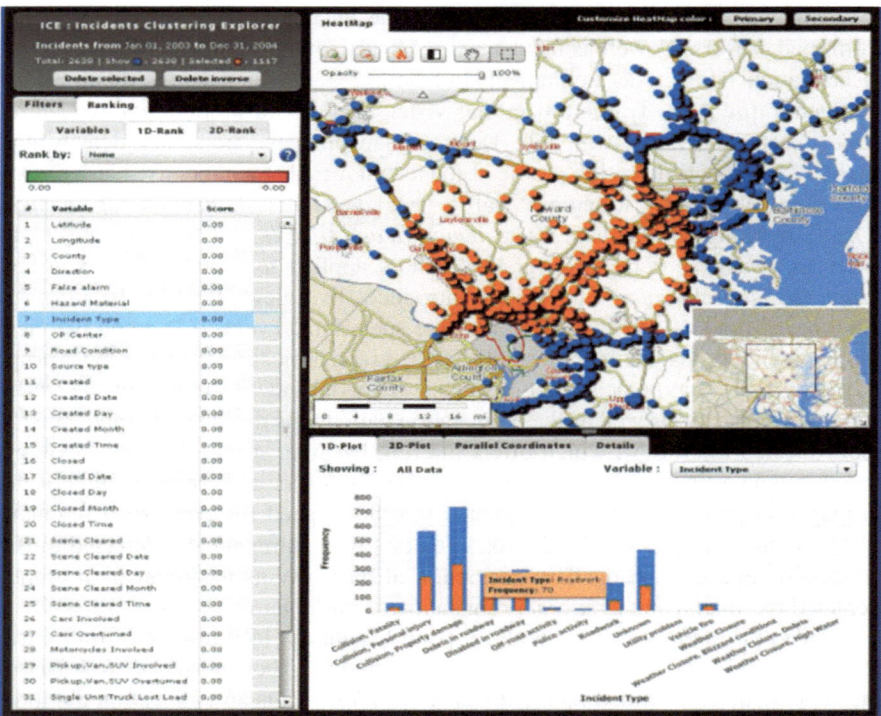

Fig. 2.3 Interactive histograms and maps are linked together. By clicking on any of the histogram bars, the incidents related to that category are highlighted in the map interface. On the contrary, the user may select a set of incidents on the map and the selected incidents are emphasized within the histogram as a highlighted subset. All rendered incidents on the map are colored in blue, and the selected incidents are represented in orange color (Adopted from [33])

immediately before a traffic incident occurs, to reveal the basic traffic flow map, and it recedes when the incident ends. They created attractive visualization in order to help traffic management controllers easily understand and access traffic and congestion data and enjoy it.

2.5   Maps as Traffic Visualization Tools

Maps are used as tools for visualization to a greater extent in traffic analysis and management as mentioned in the previous examples of applications. Denaxas et al. [37] introduced a system for assessing real-time traffic speed that retrieves the information from GPS sensors mounted in public transport fleet and renders it as color-coded segments of the road network on interactive map implementing Google Maps™ API.

A system designed by Nokia, TrafficWorks [38], collects real-time location and velocity data from GPS assisted mobile devices in vehicle. Travelers can be notified about historical, actual and future congestion levels on routes of their interest via interactive mobile map application. Chin-Hooi et al. [39] proposed a proof-of-concept system built for the traffic information system in the Klang Valley, Malaysia. The system could help drivers plan their routes and reduce traffic congestion. The diagrams demonstrate the strengths of using map-based visualization techniques to convey traffic information. It is applicable on PDA, smart phones, and computers.

Maps like Google Maps™ [40], and maps for Thailand: Traffy [41], BKKtraffic. com [42], and Longdo Map [43], supply live traffic congestion level presentation. In addition to live traffic, Google Maps™ also allows users to examine weekly traffic patterns on the map offering cumulative historical data in "Traffic at day and time" feature.

With increased precision and better reliability compared to conventional techniques, the mobile supported localization is progressively used for traffic congestion estimation service (TES) [44]. The authors of this service presented three techniques (based on mobile cell ID, based on received signal strength indicator (RSSI) and Timing Advance (TA) localization) to improve the localization precision and hence increase the trustworthiness of the estimated traffic situation. They evaluated collected statistics of all implemented techniques. The mobile assisted positioning system is then applied for TES. Server side component of the service detects and estimates congestions, and the client side is responsible for inquiring and then rendering congestion status to the user. It uses map for displaying color-coded segments of the road to the mobile subscriber where green, orange and red indicate different traffic situations, along with the average speeds of the surrounding vehicles and the distance to the congested area.

For analyzing *complex traffic trajectory data* Guo et al. [45] introduced interactive visual analytics system named Triple Perspective Visual Trajectory Analytics (TripVista) aimed for visualizing microscopic traffic trajectory data at an intersection. Using the triple perspective design philosophy (spatial, temporal and multidimensional views) as the base of the system, they designed several visualizations and convenient interactions. The system gathers traffic data at intersection through several laser scanners and some other auxiliary devices which provide more information comparing to data gained through video. The system interface is composed of three coordinated views: the traffic view for rendering spatial data, the ThemeRiver including glyphs and scatterplots for depicting temporal changes, and the parallel coordinates for multi-dimensional presentation (Fig. 2.4 adopted from [45]).

Each of the views provides appropriate users interactions. For example, selections are integrated into the traffic view. Besides, normal brushing, directional brushing is allowed to select trajectories with a certain shape. Other components of the view are ring-style sliders which are placed on top of the trajectories. The density of the trajectories is represented with histograms. Ring perimeter presented along with the histograms, from the other side, provides supplementary information regarding the traffic, for example to select the U-turn trajectories with ring sliders (Fig. 2.5 adopted from [45]).

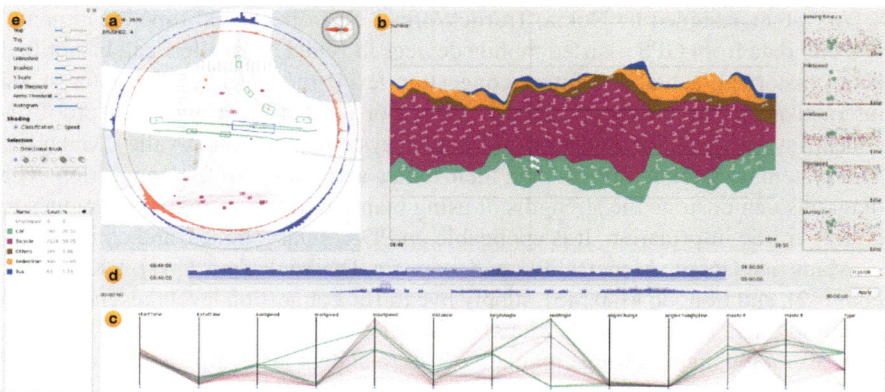

Fig. 2.4 Interface of Triple Perspective Visual Trajectory Analytics (TripVista) visualizing traffic trajectory data at a road intersection. The interface is composed of: (**a**) Spatial traffic view showing geometrical trajectory information; (**b**) Temporal views of ThemeRiver and scatterplots; (**c**) Parallel coordinates plot showing multiple properties of the multi-dimensional data; (**d**) Time sliders for two-level time range selection; (**e**) Control panel for system parameter settings and data classification. (Adopted from [45])

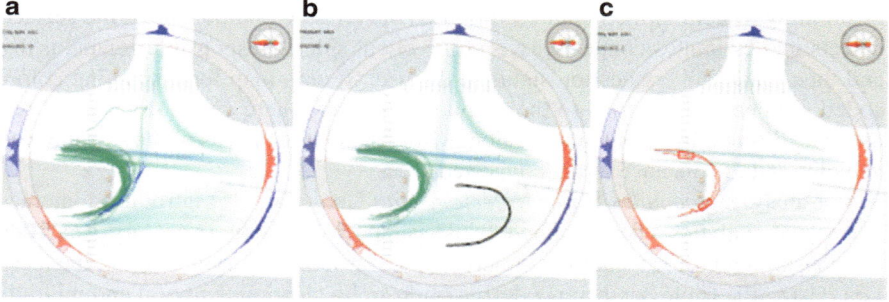

Fig. 2.5 Ring perimeter provides supplementary information regarding the traffic, enables investigation of U-turn Patterns (**a**) Select the U-turn trajectories with ring sliders; (**b**) Select the U-turn trajectories using directional brush; (**c**) Replay the scene (U-turn patterns in *red*) (Adopted from [45])

Speed is crucial for pattern examination, thus authors use another shading mode called *speed*. Here they permit the trajectory gradually to change colors while being drawn (Fig. 2.6c) adopted from [45]).

The ThemeRiver view supports user interactions such as mouse-hovering highlighting, glyph brushing and zooming. This complex design of the system for traffic data analysis is a beneficial, effective, and novel tool for the user, not only to comprehend regular traffic models (shapes), but also to reveal inappropriate behaviors.

There are maps which incorporate more *social forms of travel information* [9, 28, 46, 47]. For example digital traffic sensors are implemented with an aim to support

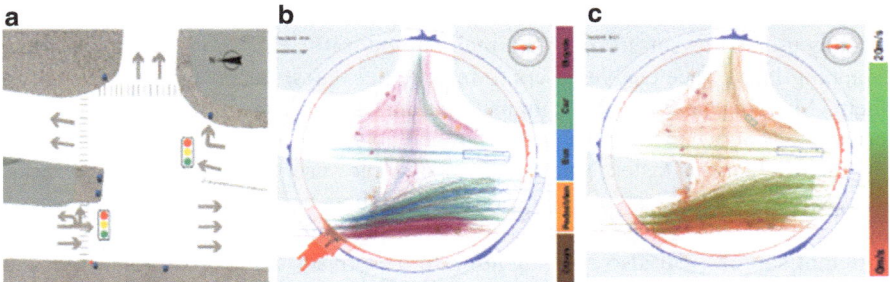

Fig. 2.6 Traffic View: (**a**) The map of the road intersection where the data is collected. The arrows on the map indicate the permitted traffic directions and the traffic light configurations; (**b**) Traffic view colored according to object types (bicycles are colored in *purple*, car in *green*, bus in *blue*, pedestrian in *orange* and others in *brown*); (**c**) Traffic view—colored according to speed variation (starting from 0 to 20 m/s, the colors representing the speed gradually change from red to green respectively) (Adopted from [45])

Table 2.1 Expenses for implementation of different sensors

Device	Cost
Vehicle loop detector	$700 for a loop
Controller	$2500
Controller cabinet	$5000
Fiber optic cable per mile	$300,000 and 10 % of the original installation cost for annual maintenance as of 1999
CCTV camera	$150–$250
Imaging-based sensing	$10–20 K per installation
Magnetic sensor-based solutions	Relatively inexpensive, they also make assumptions of traffic orderliness, unreliable for motorcycles
Pairs of inductive loop detectors (can be used to identify vehicles based on their length)	Several thousands of U.S.$ per installation-very expensive for widespread deployment and maintenance even in developed countries

travelers to share their insights and illustrate the traffic network more dynamically and proactively.

Traditional approaches for traffic information acquisition rely on a fixed sensor infrastructure. This infrastructure has high price for maintenance and deployment [30] (Table 2.1). Consequently, the traffic data coverage is restricted to main roads as well as major and vital intersections. Since this limited information impedes efficient traffic management, new approaches and applications for moderating traffic congestion are being developed.

An example of application that demonstrates the technical feasibility of obtaining sensor data and real-time traveler information regarding the road situations from active drivers is TrafficPulse [47]. It is a Volunteered geographic information (VGI) mobile application for transportation. It has a GIS platform and incorporates GPS, camera, accelerometer, and compass. The information could be shared via visualization module which supplies users with four mapping possibilities. Mapping near-by users allows users to update their current traffic condition at near real-time. Mapping latest trace enables users to review their latest traces and times spent on them. Mapping near-by public transit stations lets users to visualize the relative orientations between transit stations and their actual locations, so that they can find the nearest stations faster. Mapping events enables users to make faster decisions and help them avoid traffic congestions. The authors not only developed advanced application for smart phones that broadens the possibilities for gathering data from different sources, but also paid attention to the privacy issue, and permit users to view only their own trace when using mapping latest trace function; or to view texts like shared contents in mapping events. The system works with real-time data, but since not all smart phones have constant Internet access, it collects offline and Wi-Fi location data as well.

Hardings et al. [46] introduced iPhone application OurTravel which can be used by the users as a tool for viewing and reporting information about traffic incidents such as the type of incident (congestion, crash, road works), and exchanging personal messages or presumed delay time. They also used plots, highlight clustering around common journey end-points. The interface includes Google Maps™.

Beáta Balázsi et al. [48] introduced a prototype, Sparrow, mainly anticipated for the drivers' community, although it can be also implemented by traffic managers and analyzers. The main goal of their project is the development of a software system for broadcasting, monitoring and managing traffic data. The system is composed of a server, client application for Android based mobile phones with simple interface for reporting and displaying traffic related events, web application for the system administrators, and web application for data management and analysis. Events that can be reported cover: incidents, traffic cameras, road blockades etc. What is common with other similar applications, is that the Sparrow uses a predefined camera database, but unlike majority of that applications, it also enables users to report events in real-time. Comparing to other existing similar applications, Sparrow covers much more functionalities (map view, data clustering, filtering, visual and audio announcements), and specialized user interface which doesn't impede the driving. Besides, most approaches with camera database use only predefined databases [49], or are regions specific or limited by countries [50], which is not the case with this application. If we make a parallel with similar applications like Waze for example, which implements social networks and real time event reporting, these applications don't report only traffic related events which encumbers event filtering.

Clustering in Sparrow is classification approach implemented with an aim to properly display huge quantities of information on Google Maps™ (Fig. 2.7 adopted from [48]). To solve the issue with time-consuming standard clustering algorithms, the project uses map-reduce functions.

The client-server communication is realized with RESTful web. Other techniques that the project provides for managing different data sources are analysis and optimization via the DataMining UI. Administrators are permitted to run four types

Fig. 2.7 Map view, data clustering, filtering (administration user interface). The Sparrow software incorporates a web-based administration user interface, which permits the system administrator to analyze all the available traffic incidents and cameras on a map. The events are clustered with an aim large quantities of data to be correctly represented (Adopted from [48])

of filtering: filtering by date and reliability number, filtering duplicates, and aggregating events of the same type in a given radius.

Other examples of this type of map visualization are: INRIX Traffic [51], MIT CarTel [52] (and iCartel for iPhones). INRIX Traffic takes into consideration accidents, congestion, events, and historical traffic patterns and updates them every minute.

CarTel is a computing system intended to collect data from sensors located on mobile units such as cars and mobile phones, and then processes, delivers, and visualizes them utilizing trace, interest region or images, and hotspots for congestion depicted on the map. It is deployed on six cars, in Boston and Seattle. The system incorporates commute travel time analysis, image acquisition and their implementation as a visualization tool (Fig. 2.8 adopted from [52]).

Majority of the gathered data are geo-spatial, and the basic data segmentation abstraction is a *trace*. Traces include all sensor data collected during one trip. This library provides two classes of interfaces: (1) searching for traces using spatial queries (Fig. 2.8a) and (2) overlaying geographic characteristics on Google Maps™ for a given trace. Figure 2.8b shows the speed overlay (travel delay) as a color-coded sequence of trace's segments compatible with the vehicles' route and speed. When data are stored chronologically and user wants to find all traces for her/his travel, it is possible to "visually query" the data by means of graphically defined "interest regions" and operators (Fig. 2.8a).

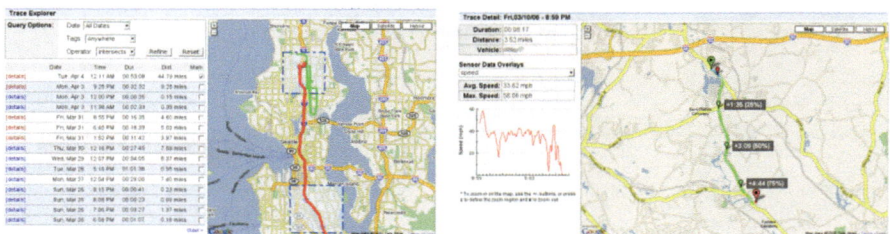

Fig. 2.8 The CarTel portal, depicting the user's: (**a**) querying for traces corresponding to user's commute and (**b**) viewing the speed overlay for one trace as a color-coded sequence of trace's segments compatible with the vehicles' route and speed (Adopted from [52]). *Map © 2015 Google Inc, used with permission. Google and the Google logo are registered trademarks of Google Inc.*

Fig. 2.9 The CarTel portal, showing a street-level view just prior to a turn (Adopted from [52]). *Map © 2015 Google Inc, used with permission. Google and the Google logo are registered trademarks of Google Inc.*

The portal offers the user more comprehensive contextual understanding of the travelled road by rendering geo-coded, street-level images of the location on Google Maps™ (Fig. 2.9 adopted from [52]). For this purpose a camera is integrated in sensor package installed in the vehicle.

Figure 2.10 (adopted from [52]) shows ten hotspots with highest congestion level. According to authors' results, in Seattle many sections of I-5 demonstrate high speed deviation during travel times. Similarly, I-93 is a key area of congestion in Boston.

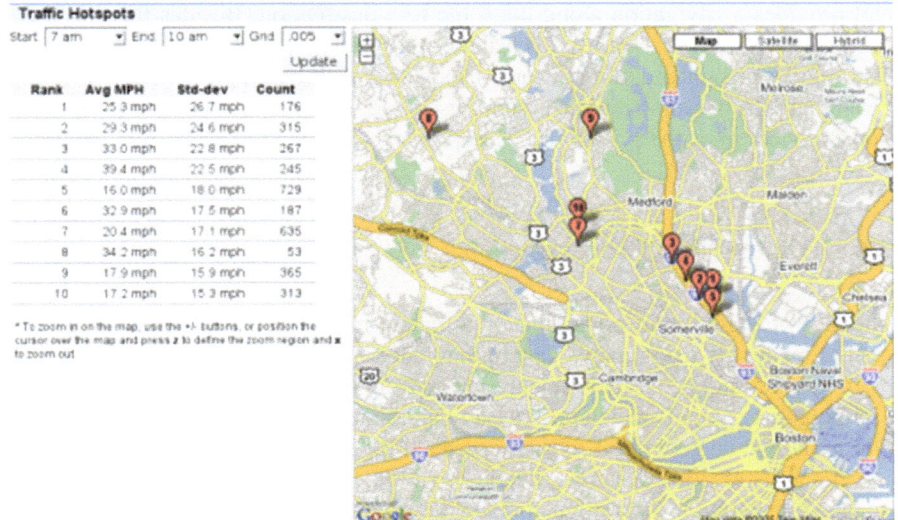

Fig. 2.10 The CarTel portal, showing ten traffic hotspots for the Boston area (Adopted from [52]). Map © 2015 Google Inc, used with permission. Google and the Google logo are registered trademarks of Google Inc.

Despite Google Maps™, Nokia Here Map is another example which offers road congestion evaluation in real time [46]. Comparing to Google Maps™, traffic information on Nokia Here is more comprehensible and better presented. The traffic incidents are displayed in list format. It provides a view of the actual congestion, road constructions, accidents and other events [53]. Furthermore, Waze is similar social mobile travel application that presents travel information generated by users to Google Maps™ [54]. It also enables travelers to communicate contextualized reports concerning congestion, delays or other journey experiences.

Volunteered geographic information (VGI) systems for transportations, such as previously mentioned CarTel [52], Google Map™ Mobile [55], Microsoft Research Nericell [56], consider that mobile devices perform *opportunistic sensing* which means they passively collect information, their state is automatically detected; and *participatory sensing* meaning human take part in data collection and participate in significant decision stages of the sensing system.

CarTel, is an example for opportunistic sensing in Boston and Seattle which monitors probe vehicles with built in GPS receivers, and transfers data back via opportunistic network connections.

NeriCell is another representative of these systems in Bangalore which applies Smartphones for monitoring road and traffic conditions. The mobile nodes are rich in sensors like GPS, Bluetooth, accelerometer, cellular radio, microphone to collect data, and there is no human interpretation involved in data collection.

Regarding its focus Waze.com is the most similar system to TrafficPulse [47], encompassing opportunistic and participatory sensing components. It records traces and provides a navigation along the route to a destination. Besides that, its user-friendly GUI allows users to contribute and visualize traffic data. Authors of TrafficPulse also intend to turn it into a social network-based platform. By offering beneficial information from the system, users are motivated to share their information and vice versa, the system can be improved by getting feedback from the users.

2.6   Digital Map Processing

The aim of digital map processing is to create raster or vector image from map images by simply using map's geographic features. One way to accomplish this is by first separating geographic-feature layers from map images using the method of grouping image pixels which possess similar colors. Other approaches use recognition of individual geographic characteristics (Fig. 2.11 adopted from [57]).

Chiang et al. [32] describe several techniques for map processing like color image segmentation, histogram thresholding, color space clustering, approaches based on region, template matching etc. Technique which is also applied in our application for congestion estimation is *histogram thresholding*. It is a technique which uses histograms of color values found in an image to identify frequently occurring values or value ranges. These values are representatives of probable color classes. *Template matching*, is knowledge-based recognition technique that could for example extract road lines, characters, and symbols from a map.

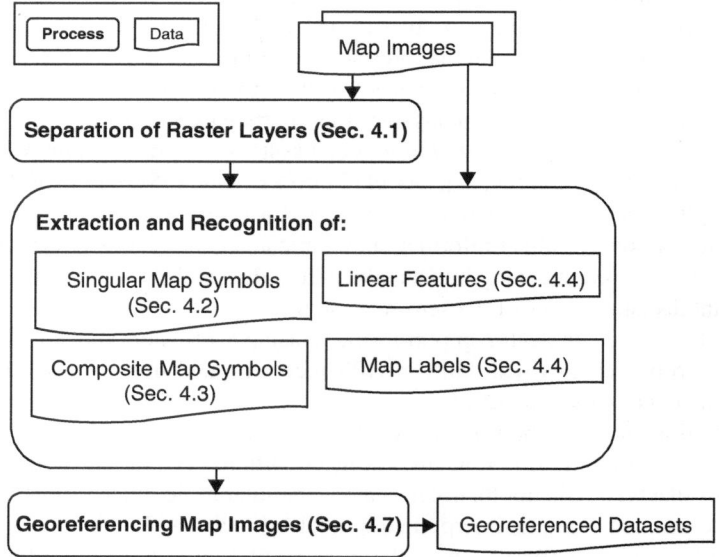

Fig. 2.11   An example map processing workflow (Adopted from [57])

Linear feature extraction and recognition techniques usually comprise separation of raster layers, and the categorization of linear features from each raster layer. Raster layers separation is usually based on histogram thresholding. One example is to extract "raw" road layers. In this context, there are some techniques that exploit more sophisticated processes for extracting the raster layers. Instead of thresholding the grayscale histogram, these methods use the RGB space and often involve user interaction. However, processes involving user interaction, such as manually determining color thresholds (Salvatore and Guitton [58]) (similar to our program for congestion estimation) or labeling colors of all line-background combinations (Khotanzad and Zink [59]) are complex, challenging, and time-demanding. Other methods for line detection prefer minimum user input or avoid relying heavily on user input (ex. manually specify the start and end points of each linear feature) in pixel-force field algorithm.

It can be concluded that processing maps is difficult and complicated. The main reason for it are the complex overlapping data which are contained in maps. Hence, minimizing the user effort in map processing and enabling users to put on their proficiency is significant. Developing efficient interactive techniques, can significantly improve human understanding of map representations and facilitate their integration in many applications.

This section reflects an intensifying need to bring together existing technologies and show how they are related and can complement each other for better transport operation analysis. It demonstrates how variety of sensors which are part of the Intelligent Transport Systems are implemented here. From the analysis of the implemented sensing it can be noticed that in the cities of the developing world, where road conditions are not good, rich sensing is critical. For example, the accelerometer is used to detect potholes and the microphone to reveal honking. Developed countries on the contrary, have high quality roads, and well managed traffic. The traffic conditions there can be defined by the velocity and volume of vehicles, more than by acceleration for instance.

Chapter 3
Visualization Tools for Traffic Congestion Estimation

3.1 Introduction

The rapid growth of urban population and number of private cars in this modern era, results in increasingly urgent transportation problem in cities throughout the world. Road traffic congestion is an omnipresent problem, which leads to delays, time loss, human stress, energy consumption, environmental pollution etc. In order to decrease traffic congestion, there is a need for simulating and optimizing traffic control and improving traffic management. There are different ways for traffic congestion monitoring and analysis such as using video monitoring and surveillance systems, or static and dynamic sensors which allow traffic management in real time. There are also other methods using non real time analysis where traffic congestion can be extracted from historical patterns of traffic congestion. The historical patterns can be gained from the stored travel time and speed data. The goal of enhancing driver convenience is achieved by providing applications based on road traffic condition that mainly identifies congestion status. This section presents a web application which uses live traffic congestion data from Google Maps™ traffic layer for real time congestion calculation. A technique utilized for estimating the level of congestion is image processing. The main objective is to provide an automated and yet interactive visualization tool for congestion analysis in real time. The aim is reducing the traffic congestion on roads which will lead to decrease in the number of accidents. The application can provide important data which can help road traffic management. Thus, it is mainly dedicated to traffic managers, operators and analysts. Nevertheless, it can be implemented also by road users. Unlike most sensor based applications, it makes quantified congestion data available even in regions with limited traffic data information.

Transport managers are often more concerned with road network overall traffic state, instead of that of some individual road links only. This application focuses on urban roads congestion analysis, with links and nodes (intersections) enclosed.

© The Author(s) 2016
N. Petrovska et al., *Innovative Web Applications for Analyzing Traffic Operations*, SpringerBriefs in Computer Science,
DOI 10.1007/978-3-319-33319-9_3

Besides the spatial aspect, the study pays attention also to the temporal dimension of road link traffic by providing stored time of the day analysis data when the congestion level was highest. The other advantage of this web application for congestion analysis is continuous provision of real-time congestion data comparing to using camera feeds, characterized with restricted view of the cameras, lack of clearness due to bad weather, and losing the connection for transferring the data. Unlike the other similar applications (for example CongestionGrid [12]), it offers more precise estimation and visualization of the congestion level and provides alerts for the user when the congestion level exceeds certain predefined by the user threshold).

The organization of this section is the following: in the second subsection the meaning of visual analytics is explained as well as its goal. The third subsection describes the system architecture. The fourth gives an insight of the visualization component and the used technology. The fifth subsection is devoted to the used methodology. The sixth subsection demonstrates the web application performance.

3.2 Visual Analytics

The main goal of visual analytics is to develop different technologies, knowledge, and techniques that combine the advantages of human and computational data processing. Visualization is a method which enables humans and computers to interact using their individual capabilities in order most effective results to be achieved. It has many applications in the field of information visualization, data mining, road traffic state analysis, geo-visualization, Geographic Information System, and many other. Geo-visualization focusses on the design and use of innovative maps and inventive functionality. There are different types of visualization: tabular representation, using graphs, plots, maps etc.

The main features of visual analysis research are: problem solving, data analysis, and/or decision-making; computational processing which involve automated procedures for data handling; human participation in the analytical process through interactive visual interfaces; support for conveying analytical results to relevant recipients.

3.3 Architecture

The application has a *stand-alone architecture*, and a web-based model. The architecture defines the software in terms of components and interactions among components. Figure 3.1 provides a view of the program architecture, drawn to emphasize a three-layer view of the software. As it can be seen from there, the layers are:

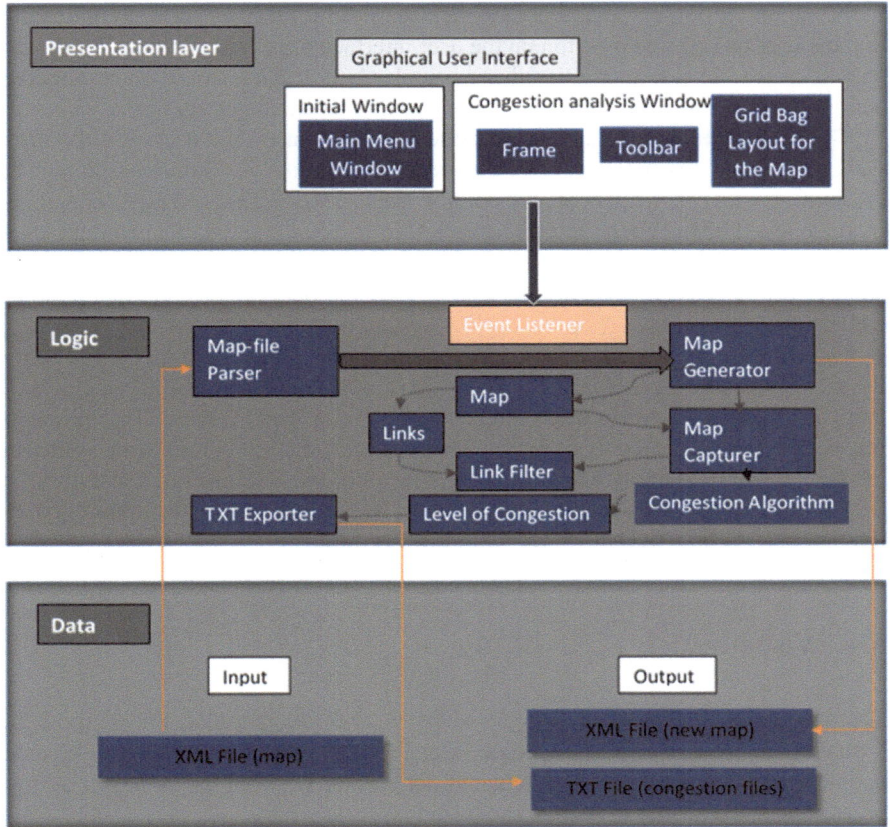

Fig. 3.1 Architecture of the software. The architecture defines the three-layer view of the software in terms of components and interactions among these components. The three layers composing the architecture are: Presentation, Logic, and Data Layer

- Presentation,
- Logic,
- Data.

The *presentation layer* is part of the software that defines what is viewable by the user and accepts user input. Actually, it is the *Graphical User Interface*. It is consisted of two basic windows: the *Main Window*, where the user may select files (maps) using browser dialogs, as well as create new maps, and a *Congestion Analysis Window*. The last has a *Frame* with a *toolbar* a *Grid Bag Layout* (map view) for the map, and the table/grid.

The *logic layer* includes a set of data *models* shown in the center of Fig. 3.1. The text boxes on the left side of the models are components that read the input file, and then export the data models to output files. The boxes on the right side of the model are components that provide the necessary processing of the input data to convert it to the summary output data. *Event listener* component receives an event indicating that the user has clicked a button on the user interface or an automatic action should be performed. The software ties this particular button click to the corresponding action (ex. parsing the new created file or opening an existing map file).

The *Map-file Parser* reads the map file name, given by the user, and the *Map Generator* generates and opens it in a browser). *Map Capturer* captures an image of the current map. Lines can be drawn on this image, covering the corresponding links on the opened map. When the *Congestion Algorithm* is initiated, the resultant level of congestion for the selected links is displayed in the scrollable table with the appropriate selected links. The user can inspect the data directly via this table as well as the map display.

The *data layer* contains *input file,* and three *output files* which are generated by the tool *Congestion Algorithm*. The input file is in XML format, which is the default map. The generated map is also represented by file in XML format. Once processing is completed, all data regarding the links' congestion is exported to a text file (in .txt format) for individual links, and for a group of links. After the threshold for high congestion is defined by the user, an output .txt format file is created containing only links with level of congestion exceeding the threshold.

3.4 Visualization

The visualization integrates Google Maps™, and image processing (Fig. 3.2). Visualization tool combines Google Map™, image-screenshot of the created zoomed map and a table with calculated congestion values. Google server allows access to the live traffic data from the traffic layer.

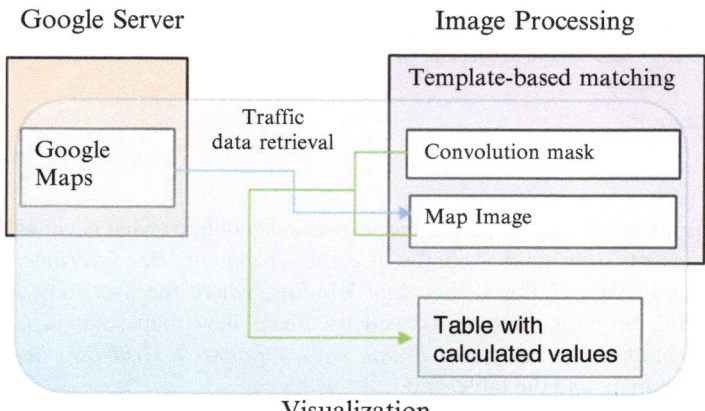

Fig. 3.2 System Architecture—the visualization integrates Google Maps™, and image processing technique. Visualization tool combines Google Map™, image-screenshot of the created zoomed map and a table with calculated congestion values

Technique used for image processing is *template matching*. It is "a technique in digital image processing for finding small parts of an image which match a template image" (Wikipedia). The method uses a convolution mask (template), tailored to a road, intersection or part of the road of the search image (the screenshot of the created map), which we want to detect. The process encompasses histogram thresholding which involves user interaction, such as he/she manually determines color thresholds and labels colors ranges. The *disadvantage* is that this technique is complex, challenging, and time-consuming.

The Web application is developed using Java programming language and computing platform. In order to run it, Java SE Runtime Environment (JRE) must be installed. It can work on PC with Windows Operating System, running standard browsers (Google Chrome, Internet Explorer or Firefox). The data regarding the congestion are acquired from the Google Maps™ Traffic Layer.

In order to obtain traffic congestion data Google utilizes the Governmental Transportation Departments' traffic sensors [60]. But the data from the infrared and laser radar are not accessible on smaller rural areas. In order to recuperate this lack of data, Google cooperates and works with the location and speed data sent by its customers who possess GPS-enabled cell phones running the Google Maps™ application, and simultaneously use the roads. Google, in order to estimate the road traffic condition, combines this provided information implementing "crowdsourcing" technique. Google has encouraged its customers who use cell phones while driving not to deactivate this feature by protecting their privacy.

3.5 Methods to Quantify Road Traffic Congestion

In this section we describe different methods to quantify current road traffic congestion based on Google Maps™ Traffic Layer. It introduces an interactive visual analytics application, Traffic Congestion Estimation Visualization Tool, for spatial and temporal exploration and analysis of traffic congestion level. Unlike the majority of previous similar works which concentrated on traffic flows for greater networks, Traffic Congestion Estimation Visualization Tool provides traffic congestion estimation on parts of a road network like links, intersections and corridors on a chosen region of the created map.

The user interface is state-of-the-art and interactive, consisted of three consecutively opened windows: the basic opening window (Main Window), HTML window comprising the new created map, and the traffic view window or *Congestion Analysis Window*.

Once the new map is created in the HTML window with the parameters chosen by the user (latitude, longitude and zoom) screenshots of Google Maps™ in the traffic view window are on-the-fly automatically generated.

The traffic view window displays the congestion information on the selected links and corridors, and selected intersections represented as polylines. The link selected to be shown on the map for analysis is displayed as a highlighted line.

The algorithm used to calculate the traffic congestion levels and to store the data into files is running at the backend. The user determines the time frequency for capturing the congestion data from the map. Simultaneously when the congestion level information is represented in the table, the congestion data are stored in text files, and may be accessed for further analysis by the user.

This visual interface allows users to geo-visualize and analyze the traffic congestion on various segments of the road network. The analysis supports road links selection and comparison of their estimated congestion.

3.6   Demonstration of the Application

To start the congestion analysis on road segments and assess road network traffic state, the first step is to create a zoomed map from the default Google Map™ (Fig. 3.3) containing the desired segments. In order to analyze road traffic congestion level from spatial perspective the user creates the new map by choosing the latitude and longitude of the center of the desired network area, and the zoom (Fig. 3.4).

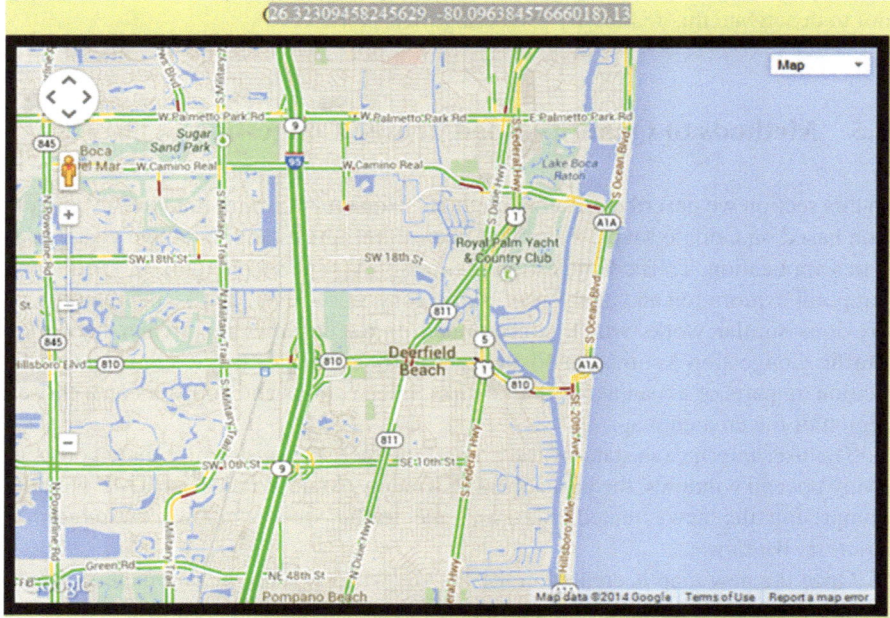

Fig. 3.3 Default Google Map™ from which a new map is created. The latitude and longitude of the chosen center of the new map, along with the zoom are displayed above the map. *Map © 2015 Google Inc, used with permission. Google and the Google logo are registered trademarks of Google Inc.*

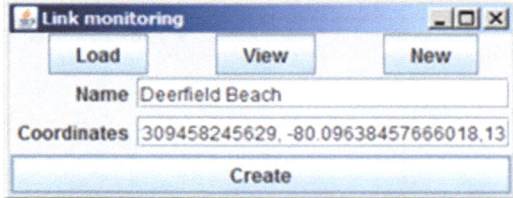

Fig. 3.4 Creating a new map from selected location on the default Google Map™. The generated map has four parameters: latitude, longitude, zoom and name as designated in the corresponding fields in the window

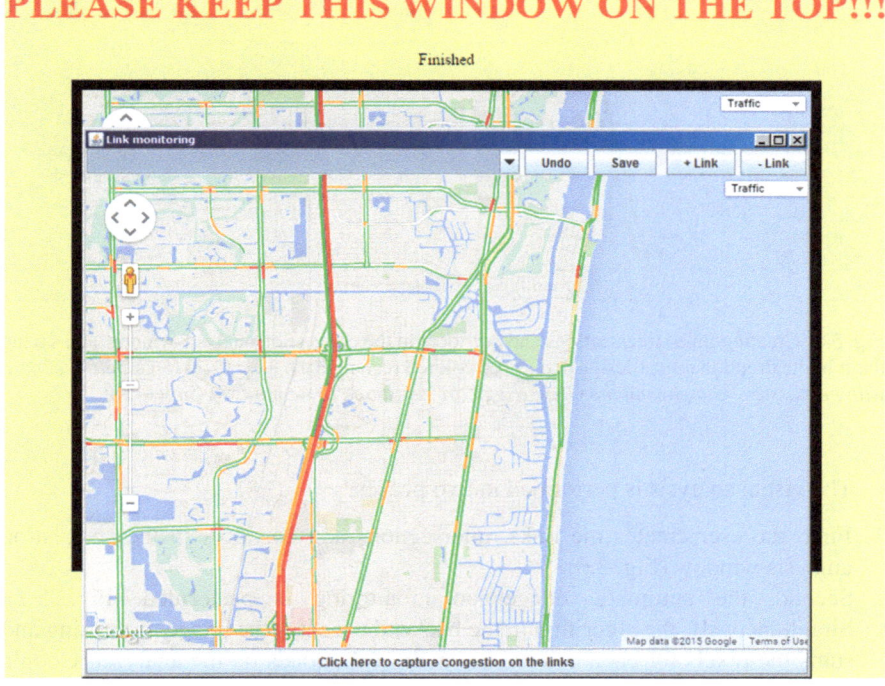

Fig. 3.5 Newly created map opens in a browser and its screenshot displays in the traffic view window i.e. Congestion Analysis Window. This window consists of buttons for creating and deleting links, dropdown list containing the created links, and a button that leads the user to a new view for traffic analysis in the same window. *Map © 2015 Google Inc, used with permission. Google and the Google logo are registered trademarks of Google Inc.*

The 20 seconds after the map is generated and displayed in a browser, the traffic view window i.e. *Congestion Analysis Window* (Fig. 3.5) is opened. It is composed of: buttons for creating, modifying and deleting links, intersections and corridors (in the toolbar); snapshot of the map (in the Grid Bag Layout); and button which opens new features for performing the congestion analysis such as table and the fields for inserting the frequency for capturing, and the congestion threshold.

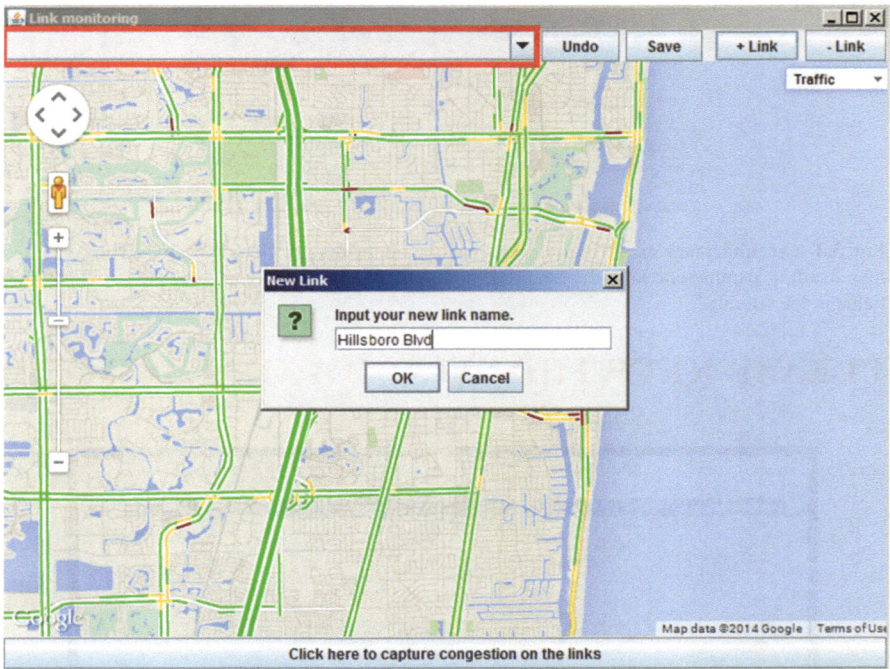

Fig. 3.6 Creating links, intersections and corridors in the congestion analysis window and saving them in the dropdown list (designated as red colored rectangular). *Map © 2015 Google Inc, used with permission. Google and the Google logo are registered trademarks of Google Inc.*

The visual analysis is performed in two phases:

1. First, the user creates the links, intersections and corridors in the congestion analysis window (Fig. 3.6).
2. Second, the automatic congestion monitoring is performed (Fig. 3.7). Simultaneously, the algorithm in the backend executes the image processing and sums the pixels for each colored segment of the roads on the analyzed Google Maps™. The estimated level of congestion is displayed in the table and expressed in percentage.

The level of congestion for each road segment is designated with letters N, M, H, and S corresponding to normal, medium, high and severe traffic congestion. Besides this, different colors are used for representing different congestion levels, such as: red color signifies high level of congestion, green indicates normal level, yellow denotes medium, while gray, severe level of congestion, as shown in Fig. 3.7. Human perception is implemented to get color classification thresholds and analyze the performance of the application. The congestion levels for every chosen link, intersection and/or corridor are recalculated with the time frequency designated by

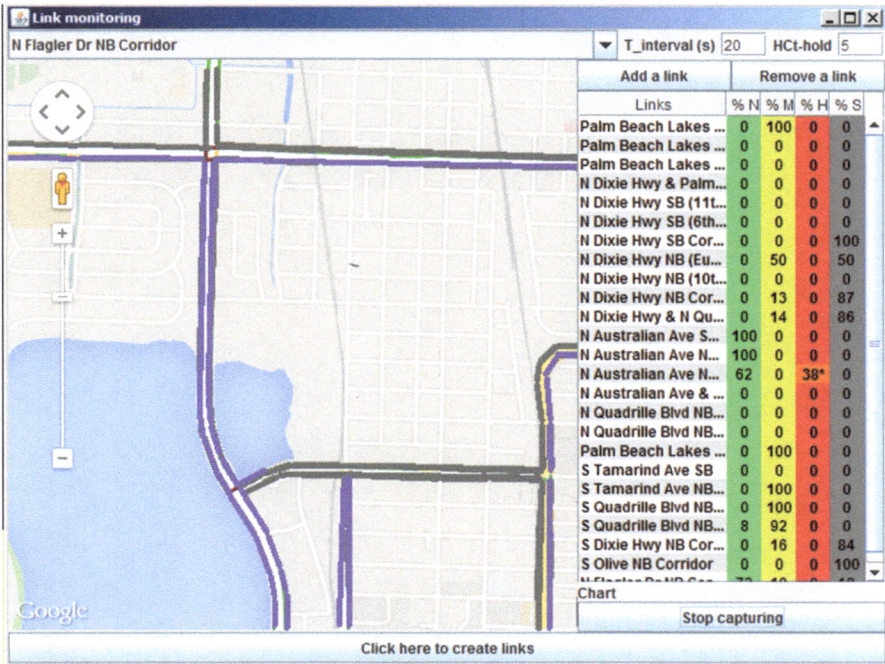

Fig. 3.7 Automatic congestion monitoring. The algorithm calculates the number of pixels for each colored segment of the road on the map and displays it in the table expressed in pixels. User selects links from the dropdown list and the selected links along with the calculated congestion values are displayed in a table. Buttons for adding or deleting a link, stopping the process of capturing data and fields for inserting time frequency for capturing and the threshold for congestion are displayed in the interface. Returning back to the previous view for creating links is allowed by the button at the bottom of the window. *Map © 2015 Google Inc, used with permission. Google and the Google logo are registered trademarks of Google Inc.*

the user and classified according to the threshold defined by her/him, for each of the road segments. Furthermore, whenever the congestion exceeds the threshold defined by the user, the application warns the user, either by alerting window or by adding an asterisk sign in the corresponding cell for the link/intersection/corridor, coloring the cell in orange.

Data obtained by the real-time traffic congestion evaluation are categorized and saved in three types of output files, containing levels of congestion for:

- the selected group of links, intersections and/or corridors,
- every link/intersection/corridor, and
- links which reached/exceeded the pre-determined congestion thresholds.

Chapter 4
Web Application for Incident Analysis

4.1 Introduction

Traffic incidents usually imply a great amount of data regarding them. Currently there is a lack of software that can be regarded as efficient incident management visualization. Most incident management systems' structures make the estimation of the actual road conditions an arduous task. It requires traffic operator to navigate through many tables, graphs, and maps to start getting a sense how to manage an incident, which is time consuming and not always correct. The emphasis is usually on data records, with less consideration given to ability for reviewing. Existing tools have proven to be not sufficiently applicable to simplify this task.

The Web application for Incidents Analysis is a visualization tool that offers incident analysts better insight of the recorded incidents. It renders historic traffic incident data from West Palm Beach incident database. The interface allows the user to plot the geospatial incident data on an interactive Google Map™. Thus, users can freely investigate regions affected by an incident and identify hot-spot areas. By using the geo-coded historical incident information, every incident can be represented with colored icon on the map. Besides investigating incidents' location, the users can filter the data by type, date, and time frame when it happened and this information is also specified via the map along with the icons. This empowers the user to gain the complete picture of the incident which is essential to expedite decision-making.

Records produced by real-time incident management systems are stored in SQL server database.

The reminder of the section is divided into several subsections. The second gives a short explanation of the nature of the incident data. The third explains the concept of the application. The fourth subsection describes the user interface and the offered user's interactivity. The section ends with information for the tools which were used for developing the web application.

© The Author(s) 2016
N. Petrovska et al., *Innovative Web Applications for Analyzing Traffic Operations*, SpringerBriefs in Computer Science,
DOI 10.1007/978-3-319-33319-9_4

4.2   Nature of Incident Data

Advanced Traffic Management Systems (ATMS) provide real-time traffic through various traffic surveillance techniques that include data from closed circuit television (CCTV), and probe vehicles, speed sensors, and other types of sensors which flow into a Transportation Management Center. The data can then be used by the traffic managers, and operators to adjust the respond to accidents.

As a source for incident data for our web application, West Palm Beach incident database was used. Incident records contain several types of data. Two most important types of data are:

- Temporal data—Time-intervals for events that have start and end times and date when the event happened.
- Geospatial data—Every incident contains latitude and longitude coordinates.

Besides these two types of data the records include specific information, such as incident ID, day of week, starting time, ending time, location, incident type, direction, lane blockage, type of the camera, signal timing changes, benefits, and intersection name. An incident location denotes a place in space on the road network where congestion has occurred due to an accident, a disabled vehicle, or any other traffic incident blocking some part of the road at that location. For our study purpose, we used approximately 60 incident records. The application is fed with these data from a file in text format. The data are stored in SQL server database and can be modified by the user of the application via the user interface.

4.3   Concept of the Application

The process for interpreting and visualizing traffic incident data starts with importing the data from a text file containing incident data for West Palm Beach. This data are saved in SQL table and displayed in the table which is part of the user interface. The application then offers the user to select an incident by its identification number. If it contains geolocation data in the table, the icon representing the incident will be displayed on the map. Otherwise the user can drag-and-drop the icon representing the incident on the map in order to retrieve geolocation from Google Server, or the user can change the existing geolocation. Other types of filtering are choosing the date when incident happened, type, time or location of the incident and displaying them on Google Maps™.

Once the geolocation is retrieved and displayed in the appropriate textboxes in the user interface, corresponding latitude and longitude can be saved in the table and SQL database.

Figure 4.1 displays the flow chart of the application for incidents analysis.

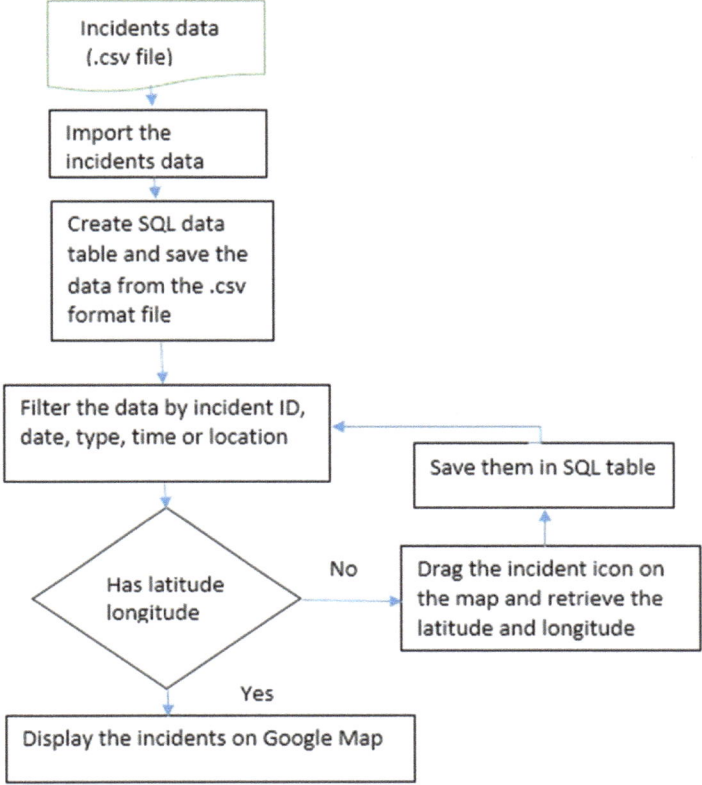

Fig. 4.1 Flow chart of the application

4.4   User Interface

The visualization tool is simple, and intuitive enough to be easily understood and provides simplification for the difficult and complex daily tasks faced by traffic incident analysts. The interface provides two types of visualization: tabular and graphical on Google Maps™ (Fig. 4.2).

Interactivity for the user and the map is provided on several ways:

- The application for incidents analysis adds the ability to drag and drop any of these incidents on the map to rearrange them as needed or to add geolocation to an incident which lacks this information (Figs. 4.3 and 4.4).
- Clicking on an icon which represents an incident will open an Info window, listing all available information regarding the specified incident.
- Multiple types of filtering are allowed for the user.
- Besides this type of interaction, zooming is also allowed.

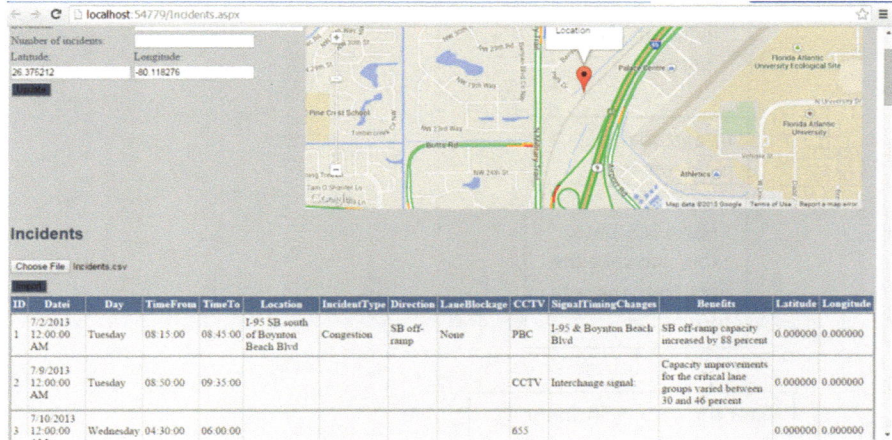

Fig. 4.2 User interface provides two types of visualization: tabular, and graphical on Google Maps™. The user imports the csv file and the data are stored in SQL table. The table contains all the information regarding the incident that are obtained from the SQL table. *Map © 2015 Google Inc, used with permission. Google and the Google logo are registered trademarks of Google Inc.*

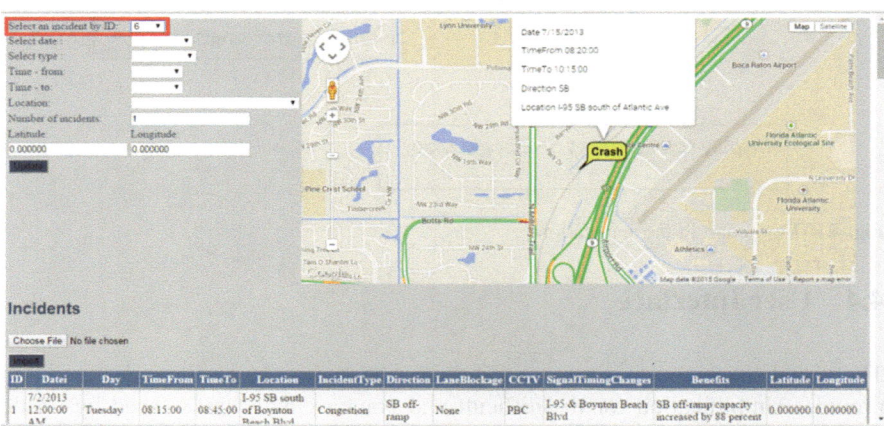

Fig. 4.3 Selecting an incident from the list (red colored rectangular). *Map © 2015 Google Inc, used with permission. Google and the Google logo are registered trademarks of Google Inc.*

When application is run for the first time, the user should first import the data from the file containing incident data in .csv format. The data from the chosen file are stored in the SQL database and the table in user's interface is populated with this data.

If the file in .csv format doesn't contain information regarding incident's location latitude and longitude, these data can be obtained by dragging the incident icon on the map, in order to retrieve the geo-coded location in the corresponding text boxes for latitude and longitude (Fig. 4.4).

Every incident is represented with a colored icon according to its type. Otherwise, if the type is not defined in the table, incident is represented with standard red colored Google Maps™ icon. By supplying visual differentiation among the types of

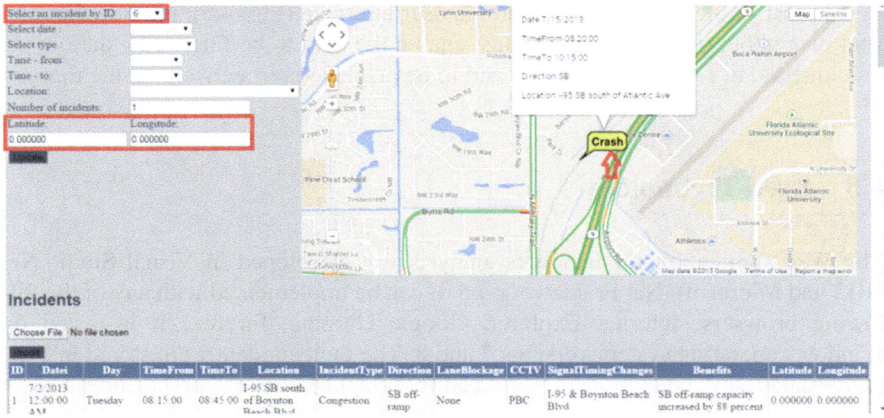

Fig. 4.4 Retrieving latitude and longitude for the selected incident from Google Maps™ server. The icon representing the incident is dragged and dropped on the map in order to provide latitude and longitude for the incident's geolocation. These data are displayed in the corresponding fields for latitude and longitude on the user interface (designated with the lower red colored rectangle on the interface). *Map © 2015 Google Inc, used with permission. Google and the Google logo are registered trademarks of Google Inc.*

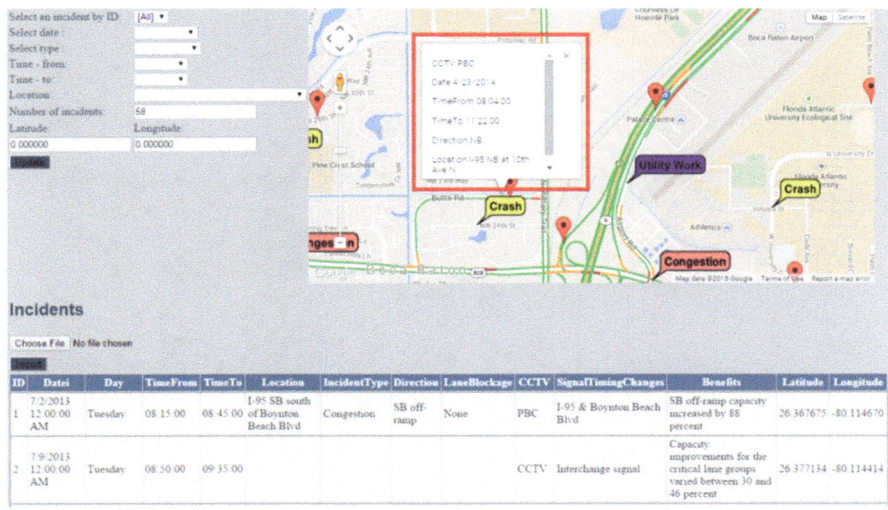

Fig. 4.5 Info window containing all the data corresponding to the incident. An Info window is displayed whenever the user clicks an icon which represents certain incident. Each incident is displayed in different color according to its type. *Map © 2015 Google Inc, used with permission. Google and the Google logo are registered trademarks of Google Inc.*

incidents, users are capable of getting faster insight of the incidents distribution within a region and find the most frequently occurred types of incidents and potential hotspots.

Figure 4.5 depicts the Info window with the extended information for the selected incident on the map.

The rest of the dropdown lists on the user interface (Fig. 4.5), provide the relevant data about their respective fields, enable the data to be filtered by date, type, time interval or incidents' location, and to be displayed respectively on the map.

4.5   Used Technology

The Web application for incidents analysis was developed in Visual Studio.Net 2013 and Microsoft. Net Framework 4.5. It can be implemented with any of the following browsers: Internet Explorer, Google Chrome, Firefox. It is tested on Windows operating system versions 7 and 8. The incident data were stored in SQL server database. The input data were read from text files in .csv or .txt format.

Chapter 5
Web Application for Measuring Travel-Based Time Signal Performance

5.1 Introduction

With increasing traffic on arterials and intersections controlled by traffic signals, the problem with delays for road users, specifically during periods of peak hours, escalates. The delays are often a consequence of not efficient operation of the traffic signals on that arterials. Many researchers proposed different analytical and heuristic methods for estimating arterial performance measures. The purpose of the Travel Time Based Signal Performance Measures application is to enable traffic operators to evaluate performance of signalized corridors by exploiting the archived upstream-link measured travel times. The web application applies the concept of Volume-Delay Functions to establish the relationship between traffic demand and measured travel times and signal control performance measures. The available signal performance measures for major through movements include: volume-to-capacity ratio, Level of Service (LOS), and number of cycles waiting. Derived measures can be graphically visualized on Google Maps™ as well as in a table on the user interface. Travel time data acquisition is performed using BlueTOAD devices or Acyclica. In order to calculate the requested performance measures, the web application exploits text files containing free flow travel times and actual travel times. Measures can be calculated on two ways, depending on which of the offered two formulas the user has chosen.

This section is organized as follows: the second subsection describes the system components, the third explains what the data component function is. Then follows the subsection which explains why the control component is needed. The fifth subsection explains the performance of the visualization component, i.e. the performance of the application, and the sixth subsection introduces the implemented technology.

© The Author(s) 2016
N. Petrovska et al., *Innovative Web Applications for Analyzing Traffic Operations*, SpringerBriefs in Computer Science,
DOI 10.1007/978-3-319-33319-9_5

5.2   System Components

Conceptually, the system consists of the following components (Fig. 5.1):

- *Data component*—collects raw data from the data provider, rearranges and maps them to system's data fields, and saves them in the database,
- *Control component*—calculates signal control performance measures based on the input values from the data component,
- *Visualization component*—the component, where the output signal control performance measures/values are displayed based on user chosen formula for calculation and input parameters.

5.3   Data Collection

In this phase, data are continually acquired and aggregated from diverse sources such as Acyclica, BlueTOAD, ATMS.now, Sensys, and CCTV cameras at every analyzed intersection. Signal timings parameters (cycle length and phase durations) are extracted from ATMS.now. Actual travel time data acquisition is performed using BlueTOAD devices or Acyclica. Acyclica generates high-resolution traffic data and offers real-time and historical data [61]. Its wireless sensors anonymously collect data at intersections and supply delay and intersection-to-intersection travel-time information which is especially beneficial for dense urban environments. As mentioned above, besides Acyclica, travel-time data are also retrieved from BlueTOAD (Bluetooth Travel-time Origination And Destination) [62], wireless identification used to connect Bluetooth™ technologies on mobile devices in vehicles, and to calculate travel time through analysis of successive detections. Traffic demands (discharge volumes and queues) are manually counted from the CCTV video recordings.

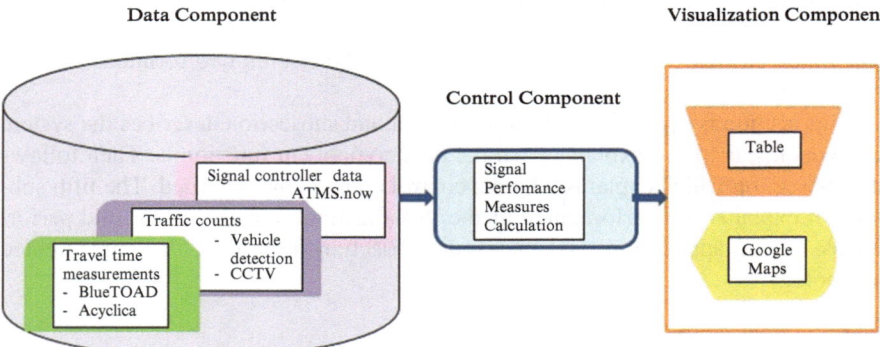

Fig. 5.1 System components. The system is composed of data component, control component and visualization component

5.4 Estimating Signal Performance

The application utilizes the concept of Volume-Delay Functions to establish the relationship between traffic demand and measured travel times, and signal control performance measures. Based on the retrieved data in the data collection phase, it calculates the signal performance measures.

Available performance measures for major through movements include: volume-to-capacity ratio, Level of Service (LOS), and number of cycles waiting on intersection. Derived measures can be graphically visualized on Google Maps™.

5.4.1 Volume-Delay Functions

There are several volume-delay formulas which were developed by different authors that can be used for evaluation of the demand/volume influence on links' travel time at macroscopic level as illustrated in Table 5.1 [68].

For the performance of this application, the first BPR formula [64] and the new derived by the LATOM[1] members [68] were implemented in the following form:

BPR formula:

$$t = t_0 * \left[1 + (1.581) * x^{2.608} \right] \qquad (5.1)$$

Newly derived VDF formula:

$$t = 0.648 * t_0 * e^{2.608} \qquad (5.2)$$

Using the results from the validation tests, it was concluded that the second equation (Eq. 5.2) is to some extent better fit for the field data.

5.4.2 Estimating Volume/Capacity Ratio

To estimate the Volume/Capacity ratio for each signal cycle, volume or traffic demand (number of vehicles on a signalized approach that want to go through a signal) was calculated as a sum of discharging volume and the number of vehicles waiting in a queue:

$$Vd = VP + VQ \qquad (5.3)$$

$$si = \left(\left(VP - (4 * NL) \right) / (GE\ T4) \right) * 3600 / NL \qquad (5.4)$$

$$ci = \left(gi / Ci \right) * si \qquad (5.5)$$

---

[1] Laboratory for Adaptive Traffic Operations and Management at Florida Atlantic University

Table 5.1 Volume-delay functions for estimating the impact of the demand/volume on link travel time at the macroscopic level

Conventional VDFs	Bureau of Public [64] Roads (BPR) function	$t = t_0 * \left[1 + a * (x)^b \right]$
	Conical VDF [65]	$t = t_0 \left(2 + \sqrt{a^2 (1-x)^2 + b^2} - a(1-x) - b \right),$ $b = (2a-1)/(2a-2), a > 1$
	Akcelik VDF [66]	$t = t_0 + 0.25T \left(x - 1 + \sqrt{(x-1)^2 + \dfrac{8bx}{d*a}} \right)$
Other VDFs	BPR2 [67]	$t = \begin{cases} t_0 + (1 + ax^b) & x \le x_{crit} \\ t_0 + (1 + ax^{b'}) & x > x_{crit} \end{cases}$
	BPR3 [67]	$t = \begin{cases} t_0 + (1 + ax^b) & x \le x_{crit} \\ t_0 + (1 + ax^b) + (x - x_{max})d & x > x_{crit} \end{cases}$
	Conical_Marginal [67]	$t = t_0 \left(2 + \dfrac{a^2 (1-x)(1-2x) + b^2}{\sqrt{a^2 (1-x)^2 + b^2}} - a(1-2x) - b \right)$
	Logistic VDF [67]	$t = t_0 + \dfrac{a}{1 + f * e^{b - d*x}}$
	Quadratic VDF [67]	$t = t_0 + a + bx + dx^2$
	Exponential VDF [67]	$t = \begin{cases} t_0 + e^{ax}/b & x \le x_{crit} \\ t_0 + e^{ax_{crit}}/b + d(x - x_{crit}) & x > x_{crit} \end{cases}$
	Inrets VDF [67]	$t = \begin{cases} t_0 \left(\dfrac{1.1 - ax}{1.1 - x} \right) & x \le x_{crit} \\ t_0 \left(\dfrac{1.1 - a}{0.1} \right) x^2 & x > x_{crit} \end{cases}$
	Lohse VDF [67]	$t = \begin{cases} t_0 (1 + x^b) & x \le x_{crit} \\ t_0 \left(1 + a(x_{crit})^b \right) + abt_0 (x_{crit})^{b-1} (x - x_{crit}) & x > x_{crit} \end{cases}$
New VDF (derived from the collected data) [68]		$t = a \cdot t_0 \cdot e^{(x*x_{max})}$

where:

Vd is traffic demand (vehicles),
VP = number of discharged vehicles on green time (vehicles),
VQ = number of vehicles in queue at the beginning of green (vehicles),
si = saturation flow rate (vehicles/lane/hour),
NL = number of lanes,
GE = effective green time (seconds),
T4 = green time used by the first four vehicles (seconds),
ci = capacity (vehicles/lane/hour),
gi = effective green time (seconds),
Ci = cycle length (seconds).

5.4.3   Estimating Level of Service

Level of Service (LOS) is a qualitative measure that explains a range of operating conditions at specific signalized intersection. HCM 2010 [18] proposes six LOS levels (A/B/C/D/E/F). HCM LOS is determined by the delay measure as follows:

Delay = Actual T.T. −Free-flow T.T.   (6)

5.4.4   Estimation of Number of Cycles to Pass Through the Signal

The number of cycles which are needed to pass through signal is calculated using the following logic: BEGIN IF (ATT/CL) < 1.0 THEN NOC < 1 cycle ELSE IF (ATT/CL) < 2.0 THEN NOC < 2 cycles ELSE IF (ATT/CL) < 3.0 THEN NOC < 3 cycles ELSE IF (ATT/CL) < 4.0 THEN NOC < 4 cycles END

where: ATT is Actual Travel Time, CL is Cycle Length, and NOC is Number of Cycles.

5.5   Visualization of VDF Measures

To allow the visualization component to efficiently utilize the extracted data, we saved the historical actual travel times in text format files. The file for free flow travel time contains five columns: column for the name of the origin intersection, a column for the destination intersection's name, and columns for free flow travel time in seconds, latitude, and longitude.

Visualization component provides user-friendly interface where the user is expected to supply alpha and beta parameters and to choose the formula which is supposed to be implemented in the calculation phase along with the values read from the free flow and actual travel time files (Fig. 5.2).

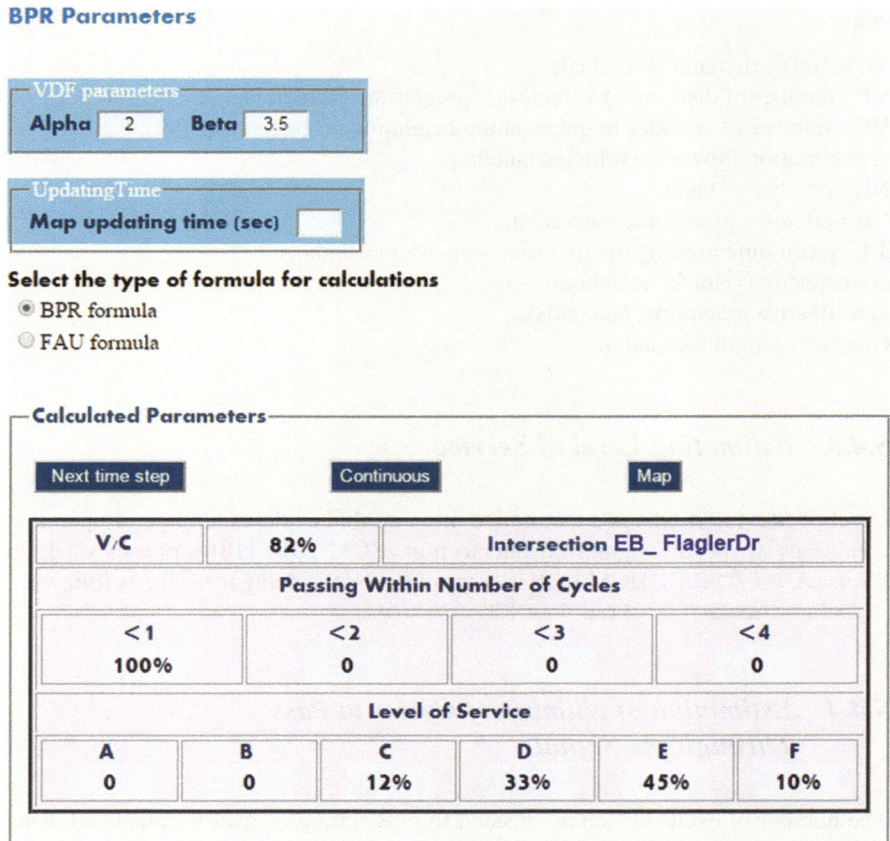

Fig. 5.2 User interface—Tabular presentation of the calculated signal performance measures. The user defines VDF parameters alpha and beta, so as to enable the calculation of the signal performance measures to be performed and chooses one of the offered formulas for calculation. The name of the intersection for which the signal performance measures are being calculated is displayed in the table. The calculated measures are displayed in %

There are two options/formulas for performing the conversion of travel time into signal performance measures: BPR or FAU equation. After the measures are calculated, the application provides three possible ways to visualize the calculated VDF measures:

- Manually, in a table, by clicking on "Next Time Step" button.
- Automatically in a table, by clicking "Continuous" button—calculated measures are displayed and updated with certain time frequency, or
- Automatically in graphical representation implementing Google Maps™ in case "Map" button is clicked.

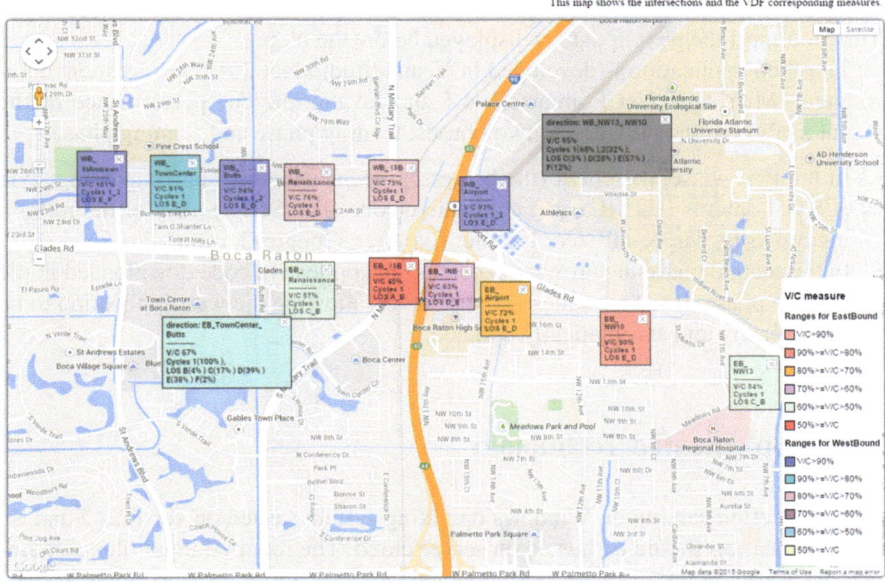

Time Left for Updating 3

Fig. 5.3 Automatic visualization of the calculated performance measures on Google Map™. Info windows, containing calculated VDF measures for each intersection are visualized on Google Maps™ simultaneously. These Info windows are automatically updated based on pre-defined time frequency in the previous window (Fig. 5.2). The time for updating is displayed in the lower left corner, below the map. On click on each Info window, a new Info window is opened containing extended information regarding the calculated measures. While the basic Info window contains only the maximum values for the number of cycles for waiting and the maximum LOS, the new Info windows contain complete information for all measures. *Map © 2015 Google Inc, used with permission. Google and the Google logo are registered trademarks of Google Inc.*

In the tabular representation besides the calculated signal performance measures for certain intersection, the name of that intersection is displayed as well (Fig. 5.2).

When selecting a "Map" as the desired output visualization, Info windows for all intersections, containing calculated VDF measures are visualized on Google Maps™ simultaneously (Fig. 5.3). These are automatically updated based on pre-defined interval in the previous window. The analysis is conducted on nine intersections for eastbound and westbound direction along Glades Road, Boca Raton, Florida. Info windows are represented in different colors, according to the range of values of the V/C parameter. The pairs, color of the Info window—V/C range, are displayed in the legend on the right side of the map.

By clicking on any Info window, additional information regarding calculated measures can be obtained. Basic Info windows contain the V/C parameter, and the maximum values for the number of cycles for waiting and the level of service expressed in %. The new Info windows, with the extended information, contain all the calculated values for the cycles, and for the levels of service, expressed in %. Time left until the next update is displayed below the map.

The Web application is developed in Visual Studio.Net 2013 development environment, Microsoft .Net Framework 4.5 and JavaScript. The performance of the visual analysis tool is divided in two phases: loading phase and running phase.

- In the first phase, the client (the browser) receives the files encompassing style (CSS), contents (HTML), and applicative code (JavaScript). Any of the following browsers can be utilized: Internet Explorer, Google Chrome, Firefox.
- In the second phase the browser executes the applicative code downloaded in the loading phase. It sends requests to the server, allowing the web application to be dynamically and automatically updated.

5.6 Implemented Technology

The application can run on Windows operating system (tested on versions 7 and 8). Web application should be hosted on server/cloud. The required input files containing the travel times used for VDF parameters calculation are text files in .csv format.

Chapter 6
Conclusion

This book introduces three applications for effective presentation of complex traffic data on web. These visualization tools can be implemented to facilitate the monitoring, interpretation, estimation and analysis of significant amount of road traffic data: congestion levels on the roads, incidents data, volume per capacity ratio, level of service, and cycles waited on intersections. Several techniques that are covered by the implementation of these web applications were described in the section which includes related work in the fields of traffic congestion, Intelligent Transport Systems, vehicular ad-hoc networks, traffic data visualization and tools, and digital map processing.

There are many advantages of these applications: improved automated visualization of traffic data, data selection and filtering, and intuitive user interfaces. Advantages of the application for congestion estimation: comparing to other tools for real-time congestion analysis, which usually require installation of high-cost devices, and very often can only monitor at fixed positions along the road (ex. by using cameras), for this application no infrastructure or additional technology is requested for performing the analysis. It uses the Google Maps™ tool, thus it is cheaper and capable of monitoring specific part of the road network. It covers data congestion in traffic data scarce information regions. Not only congestion, but quantified congestion level is offered in a user friendly interface.

Although the advantages of the technology implemented by Google to offer users the current traffic conditions in traffic layer are numerous, it has some *drawbacks*. For example:

- Active infrared and laser radar sensors don't function properly in bad weather; radar sensors can't detect stalled vehicles since they are not capable of detecting objects that aren't moving,
- The accuracy of crowdsourcing is moderate in lack of enough cell phones which are necessary for supplying reliable data for certain area.

© The Author(s) 2016
N. Petrovska et al., *Innovative Web Applications for Analyzing Traffic Operations*, SpringerBriefs in Computer Science,
DOI 10.1007/978-3-319-33319-9_6

The main goal of the application for incidents analysis was to introduce a visualization technique to emphasize the impact that traffic incidents may have on congestion. Its interface can be enriched as a future work with more descriptive visual temporal information or clustering the incidents in order hotspots to be better emphasized. Also, many incident records can be supplemented with video, and photos. These elements together with news clips or articles associated with a traffic incident could be valuable additions to this application. This application as well as the third application uses historical traffic data for visualization.

By developing these applications as efficient interactive techniques and tools, knowledge and human abilities for understanding traffic data can be better exploited.

Future work is planned to be focused on user interfaces improvement, making them more user friendly. Data mining techniques can be used for larger data volumes and more complex data analysis.

References

1. D. Levinson, "Highway Capacity and Level of Service."
2. E. D. Arnold, Jr., "Congestion on Virginia's Urban Highways," National Transportation Library, April 1988.
3. J. Zhicai, Z. Xiaoxiong, and Y. Hongwei, "Simulation Research and Implemented Effect Analysis of Variable Speed Limits on Freeway," Proceedings of the IEEE Intelligent Transportation Systems Conference, Washington, D.C., October 2004, pp. 894–898.
4. T. Thianniwet, S. Phosaard, and W. Pattara-Atikom, "Classification of Road Traffic Congestion Levels from GPS Data Using a Decision Tree Algorithm and Sliding Windows," Proceedings of the 2014 AC International Joint Conference on Pervasive and Ubiquitous Computing.
5. W. Pattara-Atikom, P Pongpaibool, and S. Thajchayapong, "Estimating Road Traffic Congestion using Vehicle Velocity," Proceedings of the 6th International Conference on ITS Telecommunications Proceedings, June 2006, pp. 1001–1004.
6. B. S. Kerner, "Tracing and Forecasting of Congested Patterns for Highway Traffic Management," Proceedings of the IEEE Intelligent Transportation Systems Conference, Oakland, CA, August 2001, pp. 88–93.
7. D. Schrank and T. Lomax. "The 2002 Urban Mobility Report," Texas Tranrpomiion Inrrimte. June 2002.
8. J. Lu and L. Cao, "Congestion Evaluation From Traffic Flow Information Based on Fuzzy Logic," IEEE Intelligent Transportation Systems, Vol. 1, 2003, pp. 50–53.
9. P. Pongpaibool, P. Tangamchit, and K. Noodwong, "Evaluation of Road Traffic Congestion Using Fuzzy Techniques," TENCON Conference, October 2007, pp. 1–4.
10. T. I. Damaiyanti, A. Imawan, and J. Kwon, "Extracting Trends of Traffic Congestion Using a NoSQL Database," Proceedings of IEEE Fourth International Conference on Big Data and Cloud Computing, December 2014, pp. 209–213.
11. A. Padiath, L. Vanajakshi, S. C. Subramanian, and H. Manda, "Prediction of Traffic Density for Congestion Analysis Under Indian Traffic Conditions," Proceedings of the 12th International IEEE Conference on Intelligent Transportation Systems, St. Louis, MO, USA, October 2009, pp. 1–6.
12. S. Pongnumkul, N. Kamsiriphiman, J. Poolsawas, and W. Amornwat, "CongestionGrid: A Temporal Visualization of Road Segment Congestion Level Data," Proceedings of 13th International Symposium on Communications and Information Technologies (ISCIT).

© The Author(s) 2016

N. Petrovska et al., *Innovative Web Applications for Analyzing Traffic Operations*, SpringerBriefs in Computer Science,
DOI 10.1007/978-3-319-33319-9

13. A. C. Diker and E. Nasibov, "Estimation of Traffic Congestion Level via FN-DBSCAN Algorithm by Using GPS Data," Proceedings of 4th International Conference on Problems of Cybernetics and Informatics, 2012.
14. https://bluetoad.trafficcast.com
15. https://ritis.org/
16. http://udottraffic.utah.gov/default.aspx
17. K.N. Balke, H.A. Charara, and R. Parker, "Development of a Traffic Signal Performance Measurement System (TSPMS)," Texas Transportation Institute, Texas A & M University System, 2005.
18. TRB, Highway Capacity Manual 2010. Retrieved from http://hcm.trb.org/, 2011.
19. H.X. Liu, W. Ma, H. Hu, X. Wu, and G. Yu , "SMART-SIGNAL: Systematic Monitoring of Arterial Road Traffic Signals", Proceedings of the 11th International IEEE Conference on Intelligent Transportation Systems Beijing, China, October 2008, pp. 1061–1066.
20. J. J. Bezuidenhout, P. Ranjitkar, and R. Dunn, "Estimating Queue Length at Signalized Intersections from Single Loop Detector Data," Proceedings of the Eastern Asia Society for Transportation Studies, Vol. 9, 2013.
21. A. Zaiat and R. J. F. Rossett, "Towards an Integrated Multimodal Transportation Dashboard," Proceedings of the IEEE 17th International Conference on Intelligent Transportation Systems (ITSC) October 2014. Qingdao, China, pp. 145–150.
22. R. Sen, B. Raman and P. Sharma, "Horn-Ok-Please," Proceedings of the 8th International Conference on Mobile Systems, Applications, and Services, pp. 137–150.
23. W. Viriyasitavat, J. M. Roldan, and O. K. Tonguz, "Accelerating the Adoption of Virtual Traffic Lights Through Policy Decisions," Proceedings of the International Conference on Connected Vehicles and Expo (ICCVE).
24. M. Ferreira, R. Fernandes, H. Conceição, W. Viriyasitavat, and O. K. Tonguz, "Self-Organized Traffic Control," Proceedings of the 7th ACM International Workshop on VehiculAr InterNETworking.
25. M. Nakamurakare, W. Viriyasitavat, and O. K. Tonguz, "A Prototype of Virtual Traffic Lights on Android-based Smartphones", Poster and Demonstration Sessions, 10th Annual IEEE Communications Society Conference on Sensor, Mesh and Ad Hoc Communications and Networks (SECON), June 2013, pp. 236–238.
26. S. Shekhar, C.T. Lu, R. Liu, C. Zhou, "Cubeview: A System for Traffic ata Visualization," Proceedings of the IEEE 5th International Conference on Intelligent Transportation Systems, September 2002, Singapore, pp. 674–678.
27. H. Piringer, M. Buchetics, and R. Benedik, "AlVis: Situation Awareness in the Surveillance of Road Tunnels," Proceedings of the IEEE Conference on Visual Analytics Science and Technology (VAST), pp. 153–162.
28. C-T. Lu, A. P. Boedihardjo, and J. Zheng, "AITVS: Advanced Interactive Traffic Visualization System," Proceedings of the 22nd International Conference on Data Engineering 2006.
29. M. L. Pack, "Wide-area, Web-based Mobility Analysis Using Probe Data," Proceedings of the 15th International IEEE Conference on Intelligent Transportation Systems Anchorage, Alaska, USA, September 2012, pp. 1682–1686.
30. J. Yoon, B. Noble, and M. Liu, "Surface Street Traffic Estimation," Proceedings of the 5th International Conference on Mobile Systems, Applications and Services, pp. 220–232.
31. Z. Wang, M. Lu, X. Yuan, J. Zhang, and H. van de Wetering "Visual Traffic Jam Analysis Based on Trajectory Data," IEEE Transactions on Visualization and Computer Graphics, December 2013, pp. 2159–2168.
32. N. Willems, H. van de Wetering, and J.van Wijk. "Visualization of Vessel Movements," Proceedings of the 11th Eurographics Conference on Visualization, pp. 959–966.
33. M. Pack, K. Wongsuphasawat, M. VanDaniker, and D. Filippova. "Ice – Visual Analytics for Transportation Incident Datasets," Proceedings of the IEEE International Conference on Information Reuse & Integration, August 2009, pp. 200–205.

34. Spotfire. http://www.spotfire.com.
35. Tableau. http://www.tableausoftware.com.
36. A. Anwar, T. Nagel, and C. Ratti, "Traffic Origins: A Simple Visualization Technique to Support Traffic Incident Analysis," Proceedings of the IEEE Pacific Visualization Symposium (PacificVis), March 2014, pp. 316–319.
37. E. Denaxas, S. Mpollas, D. Vitsios, C. Zolotas, D. G. Bleris, G. M. Spanos, and N. P. Pitsianis, "Real-time Urban Traffic Information Extraction from GPS Tracking of a Bus Fleet," Proceedings of the IEEE Symposium on Computational Intelligence in Vehicles and Transportation Systems (CIVTS), April 2013, pp. 58–63.
38. Nokia Research. TrafficWorks. Community Enhanced Traffic. http://lumiaconversations. microsoft.com/tag/nokia-traffic-works/.
39. P. Chin-Hooi, S. Kalaivani, and K Radakrishnan, "Intelligent Traffic Information System for Klang Valley, Malaysia", Proceedings of the 2008 IEEE Conference on Innovative Technologies in Intelligent Systems and Industrial Applications Multimedia University, Cyberjaya, Malaysia, July 2008, pp. 76–79.
40. Google, "Google Maps™," Internet: https://www.google.com/maps/.
41. ITS Lab, "Traffy," Internet: traffy.in.th, April 20, 2013.
42. BKKTraffic.com, "BKKTraffic.om," Internet: bkktraffic.com, April 20, 2013.
43. Longdo, "Longdo Map," Internet: map.longdo.com, April 20, 2013.
44. W. Andreas, W. Shangbo, G. H. Bruck, and J. Peter, "Traffic Congestion Estimation Service Exploiting Mobile Assisted Positioning Schemes in GSM Networks," Procedia Earth and Planetary Science, 2009, pp. 1385–1392.
45. H. Guo, Z. Wang, B. Yu, H. Zhao, and X. Yuan, "TripVista: Triple Perspective Visual Trajectory Analytics and Its Application on Microscopic Traffic Data at a Road Intersection," Proceedings of the IEEE Pacific Visualization Symposium (PacificVis), March 2011, pp. 163–170.
46. M. Harding, J. Finney, N. Davies, and M. Rouncefield, "Experiences with a Social Travel Information System," Proceedings of the ACM International Joint Conference on Pervasive and Ubiquitous Computing, September 2013, Zurich, Switzerland, pp. 173–182.
47. R-Y. Li, S. H.L. Liang, and D-W. Lee, "TrafficPulse: A Mobile GI System for Transportation," Proceedings of the First ACM SIGSPATIAL International Workshop on Mobile Geographic Information Systems, pp. 9–16.
48. B. Balázsi, O-T. Kardos, S. Ráduly, and K. Simon, "Software System for Broadcasting and Monitoring Traffic Information," Proceedings of the IEEE 12th International Symposium on Intelligent Systems and Informatics, September 2014, Subotica, Serbia.
49. http://www.ispeedcam.info/ispeedcam.html.
50. http://www.beatthetraffic.com.
51. http://inrix.com/inrix-traffic-app/
52. B. Hull, V. Bychkovsky, Y. Zhang, K. Chen, M. Goraczko, A. Miu, E. Shih, H. Balakrishnan, and S. Madden, "CarTel: A Distributed Mobile Sensor Computing System," Proceedings of the 4th International Conference on Embedded Networked Sensor Systems, pp. 125–138.
53. https://www.here.com/traffic.
54. https://www.waze.com/.
55. http://www.google.com/intl/en_us/mobile/maps/.
56. http://research.microsoft.com/en-us/projects/nericell/.
57. Y-Y. Chiang, S. Leyk, and C. A. Knoblock, "Survey of Digital Map Processing Techniques," ACM Computing Surveys, April 2014.
58. S. Salvatore and P. Guitton, "Contour Line Recognition From Scanned Topographic Maps," Proceedings of the Winter School of Computer Graphics, pp. 1–3.
59. A. Khotanzad and E. Zink, "Color Paper Map Segmentation Using Eigenvector Line-fitting," Proceedings of the IEEE Southwest Symposium on Image Analysis and Interpretation, April 1996, pp. 190–194.
60. http://smallbusiness.chron.com/google-detect-traffic-congestion-49523.html
61. Acyclica https://www.acyclica.com/

62. TrafficCast, http://www.trafficcast.com/products/view/blue-toad/
63. http://www.wired.com/2012/04/google-maps-brings-traffic-back/
64. NRC, Special Report 87: Highway Capacity Manual, 2nd edition. National Research Council, Washington, D.C., 1965.
65. H. Spiess, "Technical Note—Conical Volume-Delay Functions. Transportation Science," 1990. 24 (2), pp. 153–158.
66. R. Akçelik, "Travel Time Functions for Transport Planning Purposes: Davidson's Function, its Time Dependent Form and Alternative Travel Time Function," Australian Road Research, 1991. 21 (3).
67. PTV, VISUM User Manual Version 7.50. Retrieved from http://visiontraffic.ptvgroup.com/en-us/training-support/support/ptvvisum/, 2001.
68. A. So, A. Stevanovic, and B. Koonce, "Estimating Performance of Traffic Signals based on Link Travel Times," TRB Annual Meeting and publication in TR, January 2015.
69. N. Petrovska and A. Stevanovic, "Traffic Congestion Analysis Visualization Tool", Proceedings of the 18th IEEE International Conference on Intelligent Transportation Systems, September 2015.
70. N. Petrovska, "Software for journey planning with the public transport in Skopje", Ubiquitous Positioning Indoor Navigation and Location Based Service (UPINLBS), 2010, 14th–15th October 2010 Helsinki (Kirkkonummi), Finland, pp 1–7.
71. N. Petrovska, "Multimodal Real Time Passenger Information of the Public Transport in Skopje", ITS World Congress 2012, 22nd–26th October 2012, Vienna, Austria.